Unleaded

Unleaded

———————————————

How Changing Our Gasoline Changed Everything

CARRIE NIELSEN

Rutgers University Press

New Brunswick, Camden, and Newark, New Jersey, and London

Library of Congress Cataloging-in-Publication Data

Names: Nielsen, Carrie, author.

Title: Unleaded: how changing our gasoline changed everything / Carrie Nielsen.

Description: New Brunswick: Rutgers University Press, [2021] | Includes bibliographical
references and index.

Identifiers: LCCN 2020053273 | ISBN 9781978821002 (paperback) |
ISBN 9781978821019 (cloth) | ISBN 9781978821026 (epub) |
ISBN 9781978821033 (mobi) | ISBN 9781978821040 (pdf)

Subjects: LCSH: Lead—Environmental aspects. | Lead—Toxicology. | Lead abatement—
Government policy—United States. | Gasoline—Anti-knock and anti-knock mixtures—
Government policy—United States.

Classification: LCC TD196.L4 N54 2021 | DDC 363.738/4925610973—dc23

LC record available at https://lccn.loc.gov/2020053273

A British Cataloging-in-Publication record for this book is available from the British Library.

♾ The paper used in this publication meets the requirements of the American National
Standard for Information Sciences—Permanence of Paper for Printed Library Materials,
ANSI Z39.48-1992.

www.rutgersuniversitypress.org

Manufactured in the United States of America

For Celia and Anya and all children everywhere,
who deserve to grow up safe and healthy.

Contents

Preface

Back in 2014, long before she was awarded a $175,000 Goldman Environmental Prize for her role in exposing the Flint, Michigan, water crisis that would make headlines around the world, LeeAnne Walters was just a mom, trying to take care of her kids. The city had recently decided to save money by taking its drinking water from the nearby Flint River. LeeAnne Walters knew right away that something was wrong with the water, which came out of her tap brown and smelly, and which left her kids with skin rashes and hair loss. When her water was found to be contaminated with lead, measuring more than twenty times the legal limit, Flint officials claimed that the problem was with Walters's own plumbing, and offered to run a garden hose from a neighbor's house to provide her family with water.[1]

LeeAnne Walters contacted Miguel Del Toral at the Environmental Protection Agency (EPA). Del Toral confirmed that the Walters house, and a number of other houses in Flint, had unsafe levels of lead in their tap water. When Del Toral shared his results with others at the EPA and the Michigan Department of Environmental Quality, they tried to silence and discredit him. Del Toral would later write in an e-mail to a colleague, "I am really getting tired of the bad actors being ignored, and people trying to do the right thing are constantly being subjected to intense scrutiny as if we were doing something wrong."[2]

Miguel Del Toral introduced LeeAnne Walters to Virginia Tech engineering professor Dr. Marc Edwards. Dr. Edwards and a team of Virginia Tech students confirmed the existence of a drinking water lead crisis in Flint. Two years later, as Dr. Edwards and I sat beneath the hanging plants in his sunny university office, he denounced the government scientists and regulators who failed to uncover and properly address the Flint water crisis. "We have not figured out a way to make it good government business for scientists to be ethical," Edwards told me. He has long urged his fellow scientific researchers to be bold in the face of pushback from industry and government. Dr. Edwards himself has spent decades trying to get the problem of lead in drinking water—in Flint, in Washington, D.C., and elsewhere—taken seriously. He is still fighting.[3]

In 2018, the same week that LeeAnne Walters received the Goldman Environmental Prize, the governor of Michigan announced that Flint residents would no longer be provided with free bottled water.[4] Although tests by Marc Edwards and others have shown that tap water in Flint is safer now that it is no longer being taken from the Flint River, many residents are hesitant to trust what comes out of their taps. It makes sense to be wary when your children have been poisoned. At the height of the crisis, Dr. Mona Hanna-Attisha, a pediatrician, found that switching to Flint River water had caused the proportion of preschool-age children with elevated levels of lead in their blood to double, rising from 2.4 percent to 4.9 percent.[5] Residents of Flint are still reeling from the knowledge that their water supply poisoned one in twenty of their kids. (Note that this study, like many others, uses the term "elevated" to refer to children with blood lead levels above the current "reference level" of 5 µg/dL—a measurement that will be explained in chapter 3. Because there is no safe level of lead for children, even those whose blood lead levels are not high enough to be considered "elevated" can be harmed by lead exposure.)

Here's a part of the story that not many people know: back in the 1960s and 1970s, *all* of Flint's children had elevated levels of lead in their blood. In fact, nearly all preschool-age children in the entire United States had elevated blood lead levels. After decades of burning leaded gasoline, lead was everywhere in this country—in the air, in the dust and soil where kids played. A representative national sample of children ages five and under in the 1970s found that 99.8 percent of kids had elevated levels of lead in their blood. That last 0.2 percent is less than the margin of error for the study. Basically, the children of the United States were all lead-poisoned.[6]

For scientists like Marc Edwards and Mona Hanna-Attisha, activists like LeeAnne Walters, and conscientious government regulators like Miguel Del Toral, the fight to make the world safer and healthier for our children can sometimes seem unwinnable. In his e-mail to his colleague, Del Toral said of the situation in Flint, "It is completely stressful because it means children are being damaged and I have to put up with all of the political crap."[7] In tears, Walters told a reporter that despite the improvements that have now been made in Flint, "every time I get a call from another mother whose child is sick, it doesn't feel like a victory."[8] Recounting his years of efforts to get lead out of drinking water in Flint and elsewhere, Edwards told me that "we're in a constant battle."[9] People working to protect children from lead often have to fight for every inch of progress, and it can sometimes feel like they aren't getting anywhere.

So it's valuable to look back and see how far we've come. Those of us born in the 1960s and 1970s—Generation X—were exposed to far more lead than today's children, mainly because of the use of leaded gasoline. During those decades, scientists and doctors were hard at work measuring the scope and consequences of all that lead exposure, and getting the word out about the hazards involved. Activists were working on behalf of lead-poisoned children, raising public awareness of the problem, and fighting for policy solutions. Government bureaucrats were gathering evidence and making the case for regulations to protect children from the impacts of leaded gasoline.

Theirs was also a long, hard fight. The lead producers and oil companies fought back, working to silence or discredit anti-lead researchers and activists, and attempting to convince the public that switching to unleaded gasoline would be an economic disaster.[10] In the end, the anti-lead side won that particular battle. The amount of lead in our nation's gasoline fell by 90 percent in a single decade,[11] and by the mid-1990s, every single drop of gas pumped into cars and trucks in the United States was unleaded. The total amount of lead in the bodies of preschool children fell by more than 90 percent as well. While 100 percent of us had elevated blood lead levels back in the 1970s, today that number is less than 3 percent nationwide, and falling.[12]

This huge drop in the number of lead-poisoned kids was a big deal. Exposure to lead damages the developing brains of young children in ways that harm their learning, attention, and impulse control. Kids who have been exposed to lead have more trouble in school, a higher rate of attention deficit hyperactivity disorder (ADHD) diagnoses, and more behavior problems. They grow up more likely to be unemployed, to have an unwanted pregnancy, or even to commit a violent crime.[13] Less lead in our environment makes us all healthier and safer.

Of course, as we've seen in Flint, the fight against lead pipes—and paint, and contaminated soil—is still going on, but anti-lead activists did win the battle against leaded gasoline, and it changed everything. For those of us still fighting to make the world safer and healthier for our children, it's useful to ask what lessons we can learn from this past victory. This book is a look back at the history of leaded gasoline, the fight to switch to unleaded, and the lasting consequences of all those decades of dumping lead into our environment, and into our children.

Personally, I first learned about childhood lead exposure when I was an undergraduate at Brown University in the 1990s. I had a work-study job in a lab in the Environmental Studies Department, testing the lead levels in soil samples and drinking water samples from low-income neighborhoods around Providence, Rhode Island. I knew then that paint, pipes, and plumbing fixtures were exposing those Providence kids to far too much lead. It wasn't until years later that I saw a graph showing how much lead my own classmates and I had been exposed to as children back in the 1970s from leaded gasoline. As I learned more about the research examining the impacts of the poisoning of my entire generation, I kept thinking, "Somebody should write a book." And here we are.

In the chapters ahead, we'll look at the history of lead use and lead poisoning. As you might expect, different locations, income levels, and racial groups were exposed to very different amounts of lead over time. We'll find out about some of the shady dealings that allowed lead to be used in gasoline for so long, and meet some of the heroic scientists and activists who finally convinced the government to protect children from this toxic element. Research in neuroscience will help us to understand how exposure to lead affects the developing brains of children. We'll look at how each generation's level of childhood lead exposure affected their behavior later on, including the odds of committing violent crimes, and we'll examine how those changes in behavior and crime rates have altered our society in ways that we are still grappling with today. We'll also look at the ways in which children are still being exposed to harmful levels of lead right now. Finally, we'll learn some lessons from the history of leaded gasoline that can help us as we grapple with climate change and other challenges we're facing in the current moment, and inform our understanding of ourselves, our society, and our world.

Unleaded

1

Lead in Twentieth-Century America

▬ ▬ ▬ ▬ ▬ ▬ ▬ ▬

Back in the 1920s, near the dawn of the Automobile Age in the United States, scientists discovered that they could make gasoline into a more powerful automotive fuel by adding lead to it. Public health experts objected, calling leaded gasoline "a serious menace to the public health," "extremely poisonous," and "probably the greatest single question in the field of public health that has ever faced the American Public."[1] Nevertheless, the federal government decided to allow the use of this new fuel additive. By the 1930s, more than 90 percent of the gasoline sold nationwide was leaded, and the United States would continue to pump leaded gasoline almost exclusively for half a century.[2]

By the 1970s, a number of researchers had demonstrated the harm that exposure to lead from leaded gasoline was doing to American children. Their research showed that the lead coming out of the nation's tailpipes was damaging kids' health, their ability to learn, and even their behavior. Finally, over vehement oil company objections, the federal government decided to phase out leaded gasoline. Within a decade, the amount of lead in the nation's gasoline supply dropped by more than 90 percent, and a decade after that, every single drop of gasoline in the gas tanks of the cars and trucks on the nation's roads was unleaded. Now, in the 2020s, the United States is well into its third decade of unleaded gas.[3]

These two decisions—to allow leaded gasoline in the 1920s and to phase out leaded gasoline in the 1970s—have had far-reaching consequences. The rise and fall of leaded gasoline has affected every generation of Americans alive today, and has affected each of those generations differently. Children born during the peak lead-contamination period in the 1960s and 1970s were exposed to the highest amounts of this potent neurotoxin during their prime developmental years. Today's middle-aged adults have suffered the most from that long-ago, ill-advised choice to add lead to the nation's gasoline.

Of course, even within a single generation, exposure to lead was not equally distributed. The story of leaded gasoline is a clear example of environmental racism, demonstrating one of the many ways in which communities of color are disproportionately harmed by the impacts of industrial society. Exposure to lead does, indeed, do substantial harm. In every generation affected by leaded gasoline, and for children still exposed to lead today, the damage is serious and lasting.

Because lead exposure harms developing brains, people who were exposed to more lead as young children face a higher risk of lifelong problems with learning, memory, and impulse control. The impact of lead on brain development can have a variety of consequences throughout a person's life. For example, exposure to lead in childhood has been associated with a higher risk of dropping out of high school and a higher risk of teen pregnancy.[4] One well-documented effect of childhood lead exposure is an increased likelihood of committing a violent crime.[5] As we'll examine more fully in chapter 6, the link between lead exposure early in life and later criminal and/or violent behavior has been demonstrated by more than a dozen scientific studies, using a variety of methodologies, dating back more than twenty years.

There are many, many factors that have been shown to influence rates of violent crime—everything from economic and cultural changes to criminal justice policies to the development of new drugs and new firearms. Childhood lead exposure is just one of these many factors, and one that has often been ignored. Since the category of "violent crime" is a problematic one, with roots in the oppression of marginalized groups, any discussion of the factors influencing violent crime rates must take this history, along with ongoing bias in policing and the criminal justice system, into consideration. Differential exposure to lead is just one of the many mechanisms by which some children receive a more advantageous start in life than other children. Being exposed to lead during childhood alters brain development in ways that affect learning and behavior, and those changes can have a lifelong influence on everything from educational outcomes to criminal activity.

The baby and toddler years are the critical years when it comes to lead exposure, because that's when the largest amount of lead typically gets into the body, as babies put their grubby little fingers in their mouths, and their immature digestive systems allow most of this lead to be absorbed rather than excreted. Even more importantly, these are the years when the brain is developing most rapidly, and therefore is most susceptible to the harm that lead does to the neurons in the brain. Lead exposure later in life can still have harmful effects, but typically not as severe as the neurological damage that lead does to the developing brains of young children. So, the impact of lead exposure on particular individuals depends, primarily, on how much lead was in their environment when they were under five years old.

Small children can be exposed to lead from a number of different sources. Currently, the biggest contributor to childhood lead exposure in the United States is lead paint. As lead paint wears down over time, it produces lead-containing dust, which children encounter inside and outside their homes and other buildings. Whenever small children put their fingers in their mouths—which they do frequently—they ingest this lead and absorb it into their bodies. The use of lead in paint in the United States was highest in the 1920s, and declined slowly over the next half century. Lead as a paint additive was finally banned in 1978.[6] Unfortunately, once a building has lead paint, it always has lead paint, unless that paint is removed through a rigorous remediation process. Every building constructed before 1978 has the potential to poison kids today. Although the amount of lead from paint getting into American children has been declining for almost a century, we have a long way to go to remove this hazard. Old lead paint is the principal source of childhood lead exposure in the twenty-first century, and as we'll see in chapter 7, the fight to remove this dangerous toxin from children's homes is currently being fought in cities all over America.

Kids today are also exposed to lead from the many miles of lead pipes that carry our nation's drinking water. Lead pipe installation in the United States followed a trajectory similar to that of lead paint application—it began to wane after the 1920s but continued until the 1980s. Today, children can be exposed to lead in their drinking water when that water is not properly treated and monitored. In addition to exposure from paint and pipes, Americans have been, and continue to be, exposed to lead from industrial sites, food, cosmetics, and a number of other potential sources. Most of these sources of exposure have been declining over the past century or so, though not fast enough to protect today's children from ongoing lead poisoning. There was, however, one source of lead that followed a different trajectory—rising (and then falling) dramatically during the twentieth century, and that was the lead that was added to our nation's gasoline.

Lead poisoning is cumulative—it doesn't matter where the lead comes from; it all adds up in the body. A kid living in a house with lead paint, drinking water carried by lead pipes, and playing in soil contaminated by leaded gasoline exhaust is being impacted by all of these sources at once. My goal is not to diminish the importance of other causes of childhood lead exposure, but simply to highlight one particular cause that has affected the living generations of Americans differently. The long history of childhood lead exposure from paint and pipes has been discussed in other books. The focus of this book is on the little-told story of the dramatic impacts of the ill-advised decision to add lead to gasoline in the 1920s, and the long-awaited decision to remove it in the 1970s.

When leaded gasoline was in use in the United States, all kids breathed in this lead in the form of fine particulates coming out of the nation's tailpipes, and when those particulates settled on the ground, they became lead-contaminated dust that all kids accidentally ingested when they put their fingers in their mouths. The amount of gasoline lead that different kids were exposed to depended on a number of factors. During the era of leaded gasoline, kids living near high-traffic roads had more lead in their environments than kids living elsewhere. Urban kids were exposed to more lead than rural kids, and poor kids got a higher dose than rich kids. Because of our nation's long and shameful history of racial discrimination in housing and zoning,[7] Black kids were exposed to significantly more gasoline lead than White kids. In addition to the important influences of location, socioeconomic status, and race, the impact of leaded gasoline on any particular person alive today was determined to a large extent by when that person was born.

The amount of lead used in gasoline in the United States rose every year from its introduction in the 1920s until the beginning of the phaseout in the 1970s, then fell precipitously between the 1970s and the 1990s. Overall lead exposure in different decades of the twentieth century was determined by a combination of two patterns: slowly declining exposure to lead from paint, pipes, and other sources, along with rising and then sharply falling exposure to lead from gasoline. On average, different generations of Americans were exposed to very different levels of lead during their critical early years of rapid brain development.[8] Let's take a look at the generations who are alive today, and how each one's average lead exposure level during childhood was affected by the rise and fall of leaded gasoline.

Among those born before 1945—the Greatest Generation and the Silent Generation—some individuals were exposed to really shocking levels of lead as children. When members of these generations were growing up, the United States wasn't burning as much leaded gasoline as it would later on, but lead paint was very common, and some of that paint contained a *lot* of lead. The use of lead in American household paint peaked in the 1920s, and at that time, house paint was often as much as 50 percent lead. Eventually, other pigments became more common, and the average lead content of paint began to decline over time after the 1920s. So, Greatest Generation and Silent Generation kids whose childhood homes had older, high-lead paint in them may have been quite severely impacted. In addition, most food cans during their childhood contained lead, so they may have been exposed through their diet as well.[9]

We don't have a way to accurately estimate the average amount of lead that kids in these generations were exposed to—hardly any kids were tested for lead back then. We do know that lead exposure would have been highly variable, since it was mostly dependent on housing and diet. Kids who grew up in an older house with chipping paint may have had some of the highest levels of lead exposure of anybody alive today. However, the air and soil hadn't yet become as contaminated as they would be in later decades, so kids who managed to avoid lead paint, and didn't eat too much canned food, probably had some of the *lowest* childhood exposure levels of anybody alive today. On average, lead levels were declining during the childhoods of Greatest Generation and Silent Generation Americans, as paints began to contain less and less lead, and leaded gasoline use hadn't fully ramped up yet.

Americans born between 1946 and 1964—the Baby Boom generation—came along during the steep upward climb of leaded gasoline use.[10] Many Baby Boomers grew up in new suburban housing developments, where their parents commuted in big, gas-guzzling American cars. In addition, leisure travel was increasing, as families took those big cars on road trips all over the country. Nearly all of those cars were burning leaded gasoline. So, even as the total amount of lead in paint continued to decline, and frozen vegetables replaced canned ones in some people's kitchens, lead exposure was becoming more widespread. Every year during the childhood years of the Baby Boom generation, the United States burned more and more leaded gasoline, putting more and more lead into the air and soil. Everybody breathes the air, and all babies and toddlers put their dirt-smeared fingers in their mouths, so this contamination affected all children nationwide (though, as we've seen, some groups were affected more than others). As a generation, Boomers were certainly lead-poisoned, and "late Boomers" were even more poisoned than their older siblings.

Those of us born between 1965 and 1979 are members of Generation X. (Note that there is some controversy about the specific years assigned to each generation. I have a friend born in 1964 who swears he's Generation X, and my younger brother, born in 1979, is sometimes considered a Millennial.) When it comes to widespread childhood lead exposure in modern America, my fellow Gen Xers and I got the worst of it. We came along as the two-car family replaced the one-car family, with many of those families still driving big American cars that had gas mileage in the single digits. This country burned a *lot* of gasoline in the late 1960s and the early 1970s, and almost all of that gas was leaded.[11] So the air was full of lead, and we were all breathing it. Over time, the lead in the air settled out as a fine layer of dust on the ground; we were playing in backyards contaminated by decades of leaded gas. The very first national study of lead exposure, carefully designed to measure a representative sample of Americans, was carried out in 1976–1980, and found that a whopping 99.8 percent of children under age five had levels of lead in their blood that would now be considered "elevated." That last 0.2 percent is smaller than the margin of error of the study. When I say that all of us Generation X kids were lead-poisoned, I'm not exaggerating.[12]

After Generation X came the Millennials, born between 1980 and 2000. Millennials had the good fortune to grow up in a time when leaded gasoline was being phased out and eventually eliminated. It's true that many—far too many!—members of the Millennial generation were still impacted by all the lead paint remaining in older buildings, and other sources such as lead pipes and contaminated soil. Nevertheless, the precipitous decline of lead in the air led to a correspondingly precipitous decline of lead in the bodies of children. A new representative national survey measured lead levels in the blood of preschoolers in 1988–1994, and showed an 80 percent drop from the average levels measured in the 1970s.[13] Millennials' brains developed in a significantly less toxic environment than the previous generation's did.

Social scientists are calling the kids born since the year 2000 Generation Z. This generation—which includes my own two daughters—is growing up in an environment with the lowest levels of lead in more than a century. Most of the gasoline lead that settled onto the soil has been covered up, so the dirt that's sticking to their little fingers isn't as contaminated as it once was. The United States is also continuing to make strides in remediating old lead paint and replacing old lead-containing plumbing fixtures. That doesn't mean that the problem is solved, of course. This country has miles of lead pipe and lots of old lead paint, and we even continue to uncover new sources of childhood lead exposure. For example, some types of artificial turf playing fields are now releasing lead dust as they break down over time.[14] But kids today are growing up in a less lead-saturated environment than previous generations did. In the most recent national survey, in 2007–2010, less than 3 percent of little kids had elevated blood lead levels.[15]

The amount of lead in the gasoline used in American cars and trucks rose throughout the 1950s and 1960s, peaked in the 1970s, dropped dramatically during the 1980s, and fell to zero in the 1990s.[16] The amount of lead that each generation was exposed to depends on where we were on that curve when the members of that generation were in their critical period of early brain development. Among those of us alive today, some generations were far more lead-poisoned than other generations, on average. That difference in average lead exposure means that some generations have suffered the long-term impacts of lead on our bodies and brains, as well as on our learning and behavior, more severely than others. In addition, there was a great deal of variation within each generation. The amount of lead in each person's childhood environment was also influenced by location (urban vs. rural), by economic status, and especially by race.

As with so many benefits and challenges in American society, the dangers of lead were not, and are not, evenly distributed between White people and people of color. For much of the twentieth century, racially discriminatory laws were common. Today, racism in the United States is alive and well, and it influences everything from housing and education to health care and criminal justice. Any examination of the history and legacy of childhood lead exposure must highlight the importance of race as a factor in determining children's exposure to lead.

Black Americans have suffered disproportionately from childhood lead exposure for at least the past four generations. The biggest reason for this is housing discrimination, which had always existed in the United States, and was formalized by the Federal Housing Administration (FHA), starting in 1935. Maps of major cities across the country were created, with the neighborhoods color-coded. The neighborhoods determined to be "least desirable," commonly because they were majority Black, were outlined in red on these maps, and families living in those areas were typically denied mortgages or offered exorbitantly inflated interest rates by banks operating under the FHA guidelines. These discriminatory practices, known as "redlining," remained officially sanctioned until the 1970s.[17]

Outside of these redlined neighborhoods, many properties had deeds specifically stipulating that they could only be sold to White buyers in perpetuity. During much of the twentieth century, Black families who tried to move into predominantly White neighborhoods were threatened by their prospective neighbors with harassment and violence.[18] Some of the housing segregation we see now in the twenty-first century is a legacy of this earlier legalized discrimination, but not all of it. Even today, it has been shown that landlords discriminate against Black-sounding potential renters, and Black buyers pay more than White buyers for the same properties.[19]

All this discrimination has long forced Black families into the inner-city neighborhoods that are most likely to have lead paint and lead pipes, and these are the very same neighborhoods where high-traffic expressways are typically located. Highway construction is just one more example where the lack of political clout afforded to Black citizens has resulted in Black neighborhoods being unable to fend off the kind of harms that White neighborhoods have more successfully avoided. Environmental justice isn't just about factories and waste incinerators—living near a highway also has serious health consequences, and it's no coincidence that these highways have often been built through the middle of Black neighborhoods.[20]

The result of all that traffic that was burning leaded gasoline as it moved through Black neighborhoods—in addition to racial discrepancies in other sources of lead exposure—is that Black children have had more lead in their bodies than White children for as long as we've been measuring. In the 1970s, Black kids had about 40 percent more lead in their blood than White kids. In terms of the total amount of lead in their bodies, the switch to unleaded gas benefited Black kids even more than White kids—average blood lead levels have come down by more than 90 percent since the 1970s for both groups, and that figure represents a bigger drop for Black kids, since their levels started out higher. However, racial disparity in childhood lead exposure still persists. The most recent nationwide survey found that Black kids still have almost 40 percent more lead in their bodies than White kids on average. The numbers are lower, but the racist legacy remains.[21]

The impact of leaded gasoline use in the United States is a clear example of the existence of White privilege. I recognize that being born White has given me some very real advantages in life. Given the well-documented racial discrimination in the American housing market, I know that my parents had more options when they were first renting an apartment than they would have had if they had been a couple of color. When I was a baby, if the paint in our apartment had been in bad condition, it is likely that my parents would have been taken more seriously by the landlord or the local housing authority because of the color of their skin.

Later in my childhood, my parents owned a series of suburban houses, all of which were in predominantly White neighborhoods. They probably would not have gotten such favorable mortgage terms, or such a positive reception from their new neighbors, if they had not been White. Being able to buy homes in these "desirable" neighborhoods, and to negotiate advantageous mortgage terms, allowed my parents to accumulate household wealth over time. My husband and I are now raising our daughters in a lovely house that we were able to buy only because my parents helped us with the down payment, and that's how White privilege gets passed down from generation to generation.

It's not that all White people live like kings and queens; we have our hardships and our struggles too, but we truly have benefited from some advantages. One of those advantages relates to our ability to protect our children from the toxic effects of lead exposure. White children have been exposed to significantly less lead than Black children, on average, for as long as we've been measuring. We live in a society in which institutional racism is a measurable fact of life,[22] and combating it requires actively working to dismantle racial bias, in our institutions and in our own minds.

We've seen how leaded gasoline affected Black and White Americans—what about everybody else? Unfortunately, there isn't much historical information about childhood lead exposure among non-Black people of color. In historical lead exposure data, Black and White were typically the only racial or ethnic categories reported. Only a very few studies from the twentieth century reported results for Latinx children; and for Asian, Native American, and Pacific Islander children and other children of color, there's basically no information available. As recently as 2010, the federal government was still carrying out national lead exposure research using the racial categories of "non-Hispanic white," "non-Hispanic black," "Mexican American," and "other."[23]

Historical records do show that housing discrimination has affected all people of color in the past century, so there's evidence to suggest that some of the factors that have caused higher lead exposure for Black children have also affected other people of color. Yet, there are numerous historical and geographical differences in the experiences of different racial and ethnic groups, which likely means that there have been significant differences in childhood lead exposure among different groups as well, but we just don't have the data to know for sure.

In addition to age and race, there are a number of other demographic factors that affect how much lead you were exposed to as a child: urban versus rural, income level, region of the country, and so on. Also, there are many specific, individual differences. Did you have air conditioning when you were growing up? Opening and closing lead-painted windows in the summer is a significant source of lead dust.[24] Was there a lot of limestone around the town where you lived? Certain types of bedrock make for less-acidic drinking water, which doesn't dissolve as much lead from pipes.[25] Was it especially rainy or dry during the years when you were a toddler? The thicker the grass grows, the less lead-contaminated soil kids are exposed to.[26] However, on a national scale, all of these individual differences are swamped by the big categories of age and race.

Although there have been, and continue to be, many other important sources of lead exposure (paint, pipes, food, soil, etc.), the dramatic pattern of rising and then falling levels of lead in the blood of American preschool children in the twentieth century was driven in large part by one single source—leaded gasoline. The decision to add lead to our fuel supply, made all the way back in the 1920s, continues to have far-reaching consequences today.

2

Where the Lead
Came From

＿ ＿ ＿ ＿ ＿ ＿ ＿ ＿

From a manufacturing perspective, lead is amazing. As a metal, it's sturdy and malleable—easy to make into cooking pots and water pipes—and it doesn't rust. Mix it with tin and you get a fantastic blend for fastening other metals together. One mineral form of lead makes a dazzling white pigment for painting walls, and another mineral form makes a smoky black pigment for use in eye makeup. One organic form of lead was used as an artificial sweetener long before we had Equal or Splenda, and another organic form can greatly enhance the power of gasoline in an automobile engine. Lead-acid batteries can power a flashlight or start a car. You can use lead foil to protect the cork of a wine bottle and then drink the wine in fancy leaded crystal glasses. There are so many things lead is good for!

The problem with lead is that it's quite toxic. Throughout history, it has poisoned the people who consumed food or water from those leaden pots and pipes, ingested the pigments or the sweetener, or came into contact with air and soil contaminated by all those lead-spewing tailpipes. People working in lead-based industries always got the worst of it, and these workers suffered from a more acute version of the harm that was widespread throughout the lead-using population. Lead exposure has risen and fallen over time with trends in plumbing, paint, and gasoline, and only recently have we come to recognize that even very small amounts of lead can cause significant long-term damage.

Lead is an element with the atomic number 82, meaning that each lead atom has 82 protons, the highest atomic number of any stable element. It is the thirty-eighth most abundant element in the Earth's crust—less common than nickel and copper, but more common than silver and gold—and typically occurs in a mineral called galena. To find lead on the periodic table, you have to look for its chemical symbol, Pb, which comes from the Latin *plumbum*, the same root that gives us "plumbing" and "plumber," highlighting the long-standing use of this metal in pipes. Nearby on the periodic table are other toxic heavy metals, such as mercury, cadmium, and arsenic. These substances are elements, so they can undergo chemical reactions and form various compounds, but unlike organic toxins in the environment, they never actually break down. Not counting a tiny bit of radioactive decay, all the lead that's ever been on Earth is still here.

Humans have been using lead since at least 3500 B.C., longer than we've had written language. Lead and other metals can be extracted from certain types of rocks (called "ores") by heating up those rocks in a process called "smelting." Some metals need to be smelted at higher temperatures than others, and lead doesn't require a very high temperature at all. Even a campfire can get hot enough to extract lead from rock. So, long before humans were able to produce the extremely high temperatures necessary to smelt iron or aluminum, they could extract lead and start using it. In ancient Egypt, lead was used for a number of purposes, including fishing weights, pipes, cosmetics, and decorative objects. Lead artifacts have been found throughout the Mediterranean region, but it was the Romans who eventually started using lead in considerable quantities.[1]

When the Romans built their famous aqueducts to bring water into cities, they often used lead pipes to transport that water throughout the city. A recent study has demonstrated that those ancient pipes were releasing substantial amounts of lead into the drinking water.[2] The Romans also used lead-lined cooking pots, especially when boiling grape juice to make a syrup that was commonly used for sweetening food. When lead reacts with the acetic acid in grape juice, it forms lead acetate, which is quite sweet. Later, the Romans would make lead acetate in a crystal form and sprinkle it on food as we do with sugar. In addition, Roman women used white lead pigment on their faces to produce the pale complexions that were all the rage among the upper classes.[3]

Lead use and consumption were so common among elites during the Roman times that there has been speculation about whether lead poisoning contributed to the eventual fall of the empire.[4] As we'll discuss in chapter 5, lead poisoning damages the brain, and the behavior of some of those later Roman emperors did suggest possible brain damage. For example, Justin II, under whose reign the Roman Empire lost most of Italy, was known for biting people.[5] Lead poisoning also damages the reproductive system, which may have contributed to a low number of offspring among the Roman ruling class. While most classicists dispute the argument that lead poisoning was a primary cause of the decline of the Roman Empire, it certainly can't have helped. Well before the fall of the Roman Empire, lead was already known to be hazardous to the health of the workers who produced it, and to the lead pipe–using public as well.

Through the centuries, people continued to use lead for many purposes, and continued to express concerns about its effects on human health. In 1786, Benjamin Franklin wrote a letter to a young friend about the hazards posed by lead.[6] He mentioned a number of sources of lead exposure—distillers making rum in lead equipment, typesetters working with heated lead type, people drinking rainwater that had run across lead roofs, and so on. He also included a list of occupations whose members had been treated for lead poisoning, including plumbers, painters, and glassmakers. His letter discussed both gastrointestinal effects of lead poisoning ("the Dry Bellyach") and nervous disorders affecting the use of the hands ("the Dangles"). The letter ended with a line that environmentalists love to quote because it has proven so true: "The Opinion of this mischievous Effect from Lead, is at least above Sixty Years old; and you will observe with Concern how long a useful Truth may be known, and exist, before it is generally receiv'd and practis'd on." Indeed, it wasn't until the twentieth century that this "useful Truth" was "practis'd on" here in the country that Franklin helped to establish. In fact, Franklin wrote his letter just as the Industrial Revolution was taking hold in the United States, greatly increasing the amount of lead we would use over the course of the following century.

As in Franklin's time, by the early twentieth century, the primary concern about lead exposure related to workers who were exposed to high levels of lead in the workplace. Employees in factories that worked with lead regularly got sick, and even died, from lead poisoning. In fact, it was common practice for many of these factories to hire unskilled laborers—mostly immigrants—and then lay them off again in a matter of months, before enough lead accumulated in their bodies to make them unfit to work. A 1912 book on occupational lead poisoning in New York lists dozens of examples like that of Nathan G., who was a cobbler in his native Poland before immigrating to the United States. Nathan got a job in a paint factory as a "mixer," adding pure lead oxide powder to oil to make lead paint. This job typically involved spending all day in a room filled with lead dust. Nathan worked only three months at the factory before being hospitalized with abdominal cramps, loss of appetite, and vomiting from lead poisoning. In the early twentieth century, many industrial jobs were dangerous, even deadly, and workers were expected to accept the risks in return for a paycheck.[7]

The labor movement during the Progressive Era of the late nineteenth and early twentieth centuries pushed for workplace safety measures in many industries, including the lead industry.[8] Progress was slow and uneven, but there were finally some attempts to protect workers from exposure to high levels of lead, especially in the form of dust. Protective clothing, better ventilation, and designated areas for eating and smoking helped reduce the amount of lead that workers inhaled and ingested. Opportunities for employees to change clothes and shower before leaving for the day helped to protect factory workers' families as well. Workers in the 1930s, 1940s, and 1950s were still exposed to levels of lead that seem shocking today, but fewer of them were experiencing levels high enough to be fatal.

Factory workers weren't the only ones exposed to startlingly high levels of lead in their jobs—house painters suffered as well. The lead oxide known as "white lead" was highly valued as a pigment, for both its brightness and durability. Many professional painters mixed their own paints, and it was common to use white lead, purchased in powder form, mixed with linseed oil. The resulting paint could be more than 50 percent pure lead when dry. Painters were so dedicated to white lead paint that their trade organization convinced some state governments to implement labeling laws to ensure that the pigments they were buying were 100 percent lead. They actually demanded government-mandated labeling so that they could seek out this product, not avoid it.[9]

Mixing and applying lead paint was dangerous, but probably the most dangerous activity for painters was sanding the walls prior to painting. Working in small, poorly ventilated rooms, they gradually turned the previous coat of lead paint into a fine, airborne dust. Painters would breathe this dust while working, ingest it when they stopped for lunch, and carry it home on their clothes and skin and hair, expanding their potential exposure to twenty-four hours a day. Nervous system disorders such as "wrist drop" and "foot drop" were known to be professional hazards for painters. These two conditions are exactly what they sound like—when a person is unable to lift the hand or foot normally, due to nerve damage. The intestinal cramping and chronic constipation associated with lead poisoning were so common in the trade that they were sometimes referred to as "painter's colic."[10]

Kids were being poisoned in these lead-painted houses as well. They were exposed to lead dust when the painter sanded the walls, and over time as the lead paint slowly disintegrated into dust around them. Babies and toddlers explored their world by putting things in their mouth, as all small children do, so when chips of paint flaked off of walls and railings and windowsills, some of those flakes found their way into little bodies. The lead exposure was probably most severe when they were teething—when my own daughters were teething, they would chew on anything they could get their hands on. In a world of lead-painted furniture and toys and railings, teething kids would have ingested a lot of lead. Unfortunately, most pediatricians in the first half of the twentieth century weren't on the lookout for lead poisoning, so many cases were mistaken for other illnesses. There's no way to know how many children were harmed or killed by lead poisoning in those decades.[11] Throughout the twentieth century, and all the way up to the present time, the most severe cases of childhood lead poisoning—the ones leading to hospitalization and even death—have typically been caused by ingestion of lead paint in children's homes.

Even kids living in homes with no lead paint were being exposed to lead in other ways. People had known since Roman times that lead makes for sturdy and rust-proof pipes, and in the late 1800s, cities in the United States put that knowledge into practice enthusiastically. By the start of the twentieth century, more than two-thirds of large cities had lead pipes bringing water to their citizens. This enthusiasm for lead pipes was waning by the 1920s, due to increasing health concerns, and a number of cities had prohibited the installation of new lead pipes by 1930. However, federal guidelines promoted the use of lead pipes well into the 1950s, and many water system regulations allowed the use of lead pipes all the way into the 1980s. Chicago was still installing lead pipes in 1984.[12]

How much lead have children been exposed to in drinking water delivered through lead pipes? It is almost impossible to know. There is no good national information about how many lead pipes exist and where they are located. In addition, the chemistry of how drinking water interacts with lead pipes is quite complicated. The amount of lead entering the water from these pipes is strongly dependent on the pH and mineral content of the water. For example, the Flint water crisis of 2015 wasn't caused by the installation of new lead pipes—it happened when the drinking water plant failed to treat the drinking water properly to prevent corrosion.[13] "Corrosion" means the breakdown of the metal in the pipes—think of an old metal mailbox rusting in the rain. When the pipes started corroding, lead that had been trapped for many years was suddenly released. This combination of old lead pipes and improper corrosion control could happen at any time in most of our cities. However, when proper corrosion control practices are in place, the amount of lead coming out of lead pipes can be quite low. So we know that kids have been exposed to lead through their drinking water, but we don't know which kids or how much lead or when.

Another source of lead exposure in the twentieth century was lead solder in food cans. Solder (pronounced *sah-der*) is a mix of metals that's heated up and used as a glue to attach two pieces of metal together. This is how cans used to be made: a flat piece of metal was wrapped into a tube and its two ends soldered together, and then circles were attached to the top and bottom of the tube. Lead makes an excellent solder component—the FDA estimates that as late as 1979, 90 percent of all canned food came in lead-soldered cans. The result is not surprising—some of the lead dissolved out of the solder holding together the seams of those cans and into the food. Efforts were made during the 1970s and 1980s to improve the construction of cans and reduce this problem, and by 1986 the EPA estimated that lead exposure through diet had declined around 80 percent since the 1940s. U.S. manufacturers voluntarily eliminated all use of lead solder in 1991, and the FDA began requiring that imported food be packaged without lead solder in 1993.[14]

Lead solder wasn't the only potential source of lead in the food supply. Lead arsenate was a common pesticide used on fruits and other crops early in the twentieth century, causing many foods to be contaminated with lead pesticide residue. Although its use declined dramatically after DDT was developed in 1947, lead arsenate use wasn't officially banned until 1988. Because lead is an element, it doesn't break down over time like many organic pesticides do, so lead residue can still be found in the soils of orchards treated with lead arsenate even many decades later. In fact, experts have recommended against eating mushrooms grown in abandoned apple orchards because they can contain dangerous levels of lead.[15] It's impossible to know for sure how much lead entered the bodies of children from lead-soldered cans and lead pesticide residues over time, but the evidence suggests that lead exposure through food declined steadily during the second half of the twentieth century.

American children in the twentieth century faced a number of other possible routes of exposure as well. (Pencils, however, were never a threat—the "lead" in pencils is, and has always been, graphite, which is actually a form of carbon that was historically referred to as "black lead." So no worries, all you pencil nibblers.[16]) Lead can still be found in certain imported toys and candy, folk remedies, and jewelry. Even certain types of artificial turf playing fields have been found to contain lead and to release lead-contaminated dust as they break down over time.[17]

But the primary focus of this book isn't on paint or pipes, food or toys—it's on leaded gasoline. There are two reasons for this focus on the lead that we once put into our gas tanks and spewed out of our tailpipes. For one thing, leaded gasoline affected *everybody*. It didn't matter what kind of paint was in your house, or what kind of plumbing your city used, or whether you grew your own vegetables—everybody breathes the air, and all children end up ingesting some dust and dirt. During the years when we were using leaded gasoline, every single child in America was exposed to that lead. Of course, that exposure was never evenly distributed—race was a major factor in determining different kids' level of exposure, and so did location and income. Some groups were exposed more than others, but nobody was spared.

The second reason for this book's focus on leaded gasoline is that it's the factor that best accounts for high levels of lead exposure in the 1960s and early 1970s, and the sudden, dramatic drop in lead exposure after the mid-1970s. The amount of lead in our paint and our food had already been declining slowly for many decades before 1970. The amount of lead in automotive gasoline, on the other hand, rose dramatically in the 1940s, 1950s, and 1960s, peaked in the 1970s, and then dropped to zero within twenty years. This book tells the story of a generation born in the 1960s and 1970s that was exposed to significantly more lead than the generations that came before and after them, and that story is primarily a story of the rise and fall of leaded gasoline. It's a story that begins in the 1920s.

The first mass-produced automobiles were made at the very beginning of the twentieth century, but sales didn't really take off until a decade later, when Charles Kettering invented the electric starter. Before that, it took strength and bravery to start a car—the crank was difficult to turn, and if the car backfired it could break your arm, or worse.[18] By the early 1920s, automobiles were becoming more common in the United States, and consumers were demanding faster, more powerful cars.

Unfortunately, these powerful engines suffered from a problem called "knocking." As you may know, cars are powered by tiny explosions. Inside an automobile engine, a mixture of gasoline and air is injected into a cylindrical chamber, and a piston forces its way in, compressing the gas-and-air mixture inside the cylinder. Then, at just the right moment, the spark plug produces a spark that ignites the gas, creating a small bang that forces the piston back out of the cylinder. It's the force of that explosion pushing on the piston that turns the crankshaft that makes the wheels go around. The farther into the cylinder the piston goes before the explosion, the farther back it gets pushed, and the more force it has to turn the crankshaft. More compression means more power. That's why a car with a higher "compression ratio" goes faster.

Here's the problem: when a gas gets compressed, it heats up. You've experienced the opposite of this if you've ever held down the nozzle on an aerosol spray can for a while. The can gets cold, because the contents are getting uncompressed as they come out of the can. What's happening inside a car's engine is that the gases are getting more compressed, so they're getting hotter. Eventually, if the gas-and-air mixture heats up enough, it will explode on its own, without any help from the spark plug. That's not good.

When the spark plug ignites the fuel, it does so in a very particular way. Ignition starts at one end of the cylinder, and happens when the piston is in exactly the right position. This results in a controlled little explosion, perfectly calibrated to push the piston back down the cylinder and power the car. Uncontrolled detonation, however, can happen at the wrong time and in the wrong location. These spontaneous explosions are called "knocking," and they are bad news. Unlike the controlled explosions produced by the spark plugs, knocking explosions don't push on the pistons properly, so knocking significantly reduces the power that the engine gets from burning the fuel, and over time, these uncontrolled explosions can actually destroy the engine.

Some fuels are more prone to exploding when they get compressed than other fuels are. A fuel's octane is a measure of its ability to resist exploding under pressure, and so prevent knocking. It's called "octane" because it's measured in comparison to a particular hydrocarbon, iso-octane.[19] The type of gas commonly used in the early twentieth century had quite low octane, so those cars couldn't have a very high compression ratio—that is, the piston couldn't squeeze the gas-and-air mixture very much before it was ignited. That's part of why a Model T had a top speed of 45 mph.[20]

If people wanted faster, more powerful, more efficient cars, they had to figure out a way to have more compression without knocking, which required higher-octane fuel. In the early twentieth century, a lot of experimentation was going on with different types of fuels and different types of engines. Sun Oil (now known as Sunoco) was selling a type of naturally high-octane gas from California petroleum, but this was relatively expensive and in short supply. Others were working on a chemical process called "thermal cracking" that heats up the hydrocarbons in petroleum and alters the molecules in a way that raises the octane of the resulting fuel. Implementing this process on a large scale would have required oil companies to install expensive new equipment in all of their refineries, which wasn't a popular option among oil company executives and shareholders.[21]

There was already one other widely known method for raising the octane of fuel—adding ethanol. Ethanol (ordinary alcohol, the same stuff found in a vodka martini) is much better than gasoline at withstanding compression without spontaneously exploding. In fact, for a long time, there was a great deal of interest in using ethanol as a fuel all by itself. The very first internal combustion engine ever invented, in 1826, ran on a combination of alcohol and turpentine. The first four-stroke internal combustion engine (the type in our cars today) ran on alcohol, and so did the first engine that Henry Ford ever built. In fact, some early Ford Model A cars had a knob on the dashboard that allowed the driver to switch between gasoline and ethanol. By 1920, research was under way into the use of various combinations of alcohol and petroleum fuels.[22] Unfortunately, both thermal cracking and ethanol mixing would soon be overtaken by a brand new method of raising octane and preventing knocking.

General Motors (GM) had hired Charles Kettering—the same guy who invented the electric starter and made cars drivable by just about anybody—to work on the problem of engine knocking. GM intended to outsell Ford and its other competitors by producing bigger, faster, more powerful automobiles, but it couldn't do that until it solved the knocking problem.[23] Kettering assigned Thomas Midgley to work on the project. It was Midgley who figured out that the problem was the uncontrolled explosion of compressed fuel, and he developed the "octane" scale that we still use to measure a fuel's resistance to these explosions. At one point, Midgley, like many of his contemporaries, was working with alcohol—in 1920, he filed a patent application for a mixture of cracked petroleum and ethanol. He also tested many other compounds, but the only ones that worked were highly expensive, corrosive, or—in one case—just too smelly. Poor Mrs. Midgley made her husband sleep on the couch while the lab was working on this compound, which smelled of onions and garlic, a scent that resisted scrubbing and lingered for days.[24]

Eventually, Midgley and his associates took out a periodic table and highlighted all the elements that they had found to have some ability to reduce knocking. The resulting pattern pointed them toward another promising element—lead. Back in 1853, a German chemist had developed an organic form of lead called tetraethyl lead (TEL), which consists of a lead atom attached to four ethyl groups. (An ethyl group is just a particular arrangement of hydrogen and carbon.) Tetraethyl lead dissolves easily in gasoline, and Midgley quickly determined that a small amount of TEL increased the octane of the gas dramatically.[25] This, Midgley told a collaborator while holding a test tube of TEL, is "really the answer to the whole problem."[26]

Just like that, General Motors had an antiknock additive that was cheap, effective, and—most importantly from a corporate perspective—patentable. Adding TEL to gasoline would allow GM to double the compression ratios of its engines, allowing them to build the bigger, faster, more powerful cars they had been hoping for.[27] General Motors and Standard Oil eventually got together and created a company called the Ethyl Corporation to sell this new product. In a stroke of marketing genius, they called their new gasoline additive "ethyl," carefully leaving out any reference to lead, which was already associated with toxicity in many people's minds. The Ethyl Corporation hired the DuPont company to manufacture TEL, and they got to work scaling up production.[28]

From the beginning, some of those involved in the development of TEL had concerns about its safety. In 1922, DuPont was run by two brothers, and one brother wrote to the other brother that their new product, TEL, was "very poisonous if absorbed through the skin, resulting in lead poisoning almost immediately." One of Thomas Midgley's co-workers, Tabby Boyd, later said that "from the outset it was appreciated that putting tetraethyl lead into gasoline might possibly introduce a health hazard. The first opinions of the doctors who were consulted were full of such frightening phrases as 'grave fears,' 'distinct risk,' 'widespread lead poisoning.' The source of the possible hazard to health thought of at first was not so much that from the tetraethyl lead itself as that from finely divided lead dust in engine exhaust."[29] In addition, Kettering's lab received letters from numerous public health and toxicology experts expressing their concerns about TEL. Midgley himself suffered from lead poisoning in 1923 and went to recuperate in Florida.[30]

The federal government was also hearing from concerned scientists and public health experts. In October 1922, chemistry professor William Mansfield Clark wrote to A. M. Stimson, who was the assistant surgeon general at the Public Health Service, that TEL posed "a serious menace to the public health." Clark correctly predicted that "on busy thoroughfares it is highly probable that the lead oxide dust will remain in the lower stratum."[31] After reading Clark's letter,

Stimson concluded that "the possibilities of a real health menace do exist in the use of such a fuel and it is deemed advisable that the [Public Health] Service be provided with some experimental evidence tending to support this opinion." He requested that the Division of Chemistry and Pharmacology investigate the issue. However, the director of the division declined, claiming that such trials would be too time-consuming, and suggested that the lead industry itself could provide the needed information about potential health risks.[32]

A month later, the surgeon general wrote to one of the DuPont brothers: "Inasmuch as it is understood that when employed in gasoline engines, this substance will add a finely divided and nondiffusible form of lead to exhaust gases, and furthermore, since lead poisoning in human beings is of the cumulative type resulting frequently from the daily intake of minute quantities, it seems pertinent to inquire whether there might not be a decided health hazard associated with the extensive use of lead tetraethyl in engines."[33] Already, it was recognized that the lead that comes out of a tailpipe lingers in the environment, and will result in ongoing chronic lead exposure over time; it was indeed "pertinent to inquire" about the "decided health hazard" this would create. Thomas Midgley wrote back on behalf of DuPont, and while he admitted that "no actual experimental data has been taken," nevertheless he argued that "the average street will probably be so free from lead that it will be impossible to detect it or its absorption."[34]

The Public Health Service had declined to study the issue, but the federal Bureau of Mines agreed to carry out a study funded by General Motors itself. The Bureau of Mines was widely known to be a pro-industry agency, and that reputation was certainly borne out by the terms that they approved for their study of tetraethyl lead. They agreed to refer to the compound only as "ethyl," leaving out any mention of lead. The study would focus primarily on the safety of the workers producing TEL, and not on issues relating to the general population and the use of TEL in gasoline. In addition, GM reserved the right to approve any results before they were published, effectively giving themselves veto power over the entire study. Needless to say, this was not the kind of unbiased, wide-ranging study that public health experts had been looking for. Dr. Yandell Henderson, a professor of applied physiology at Yale University and a chief scientific adviser to the Workers' Health Bureau, was asked to assist the Bureau of Mines with their study, and he wrote back to state his terms: "I should be glad to investigate the physiological and sanitary questions concerned, but only on the assumption that so terrible a poison as tetra-ethyl lead should not be generally introduced until absolute proof was available that no danger to the public would be involved." Dr. Henderson was not invited to participate in the study after all.[35]

While the Bureau of Mines study was under way, production and marketing of "ethyl" began in earnest, and the very first gallon of leaded gasoline was pumped on February 1, 1923, at a gas station in Dayton, Ohio, owned by a friend of Charles Kettering's. Originally, TEL was sold as a dark-red fluid that the gas station attendant would mix into the gasoline right there at the pump. Ethyl gas was a big hit—drivers found that it did indeed stop knocking, as well as increasing the power of their engines and improving their fuel efficiency. In some of the newer cars, with their high-compression engines, the increased octane of leaded gasoline allowed people to drive as much as 50 percent farther on a gallon of gas. Within eighteen months, Ethyl gas was being sold at 10,000 gas stations in twenty-seven states.[36] Leaded gasoline was a big success at the pumps, but in the factories producing the TEL, things started to go catastrophically wrong.[37]

DuPont had managed to cover up the earlier deaths of several workers at two of its TEL plants, but a series of grisly deaths in 1924 at their Elizabeth, New Jersey, plant made headlines. Five workers died and thirty-five others experienced the convulsions and hallucinations associated with severe lead poisoning. One employee had to be subdued by three strong men before he could be taken away in a straitjacket. Lurid newspaper coverage of the violent insanity of the poisoned workers—the fumes in the factory were referred to as "looney gas"—raised immediate public concern. Health experts took the opportunity to point out that, with leaded gasoline already being pumped into gas tanks around the country, it wasn't just industrial workers who were at risk. Dr. Yandell Henderson, the Yale physiologist who had declined to participate in the biased Bureau of Mines study, told the *New York Times*, "This is probably the greatest single question in the field of public health that has ever faced the American Public."[38] Soon, New York City, Philadelphia, Pittsburgh, and the state of New Jersey had banned all use of TEL in gasoline.[39]

The day after the fifth worker died at the Elizabeth, New Jersey, plant, the results of the questionable Bureau of Mines study were released, and unsurprisingly, the study found no health hazards associated with "ethyl." This result was criticized by numerous scientists, public health experts, and labor activists, and at least one competitor of General Motors. For example, Dr. Alice Hamilton, a physician whose pioneering work in the field of occupational health had made her a leading expert on lead poisoning, strongly condemned the study's flawed experimental conditions and questionable funding source.[40] Dr. Hamilton argued for the "desirability of having an investigation made by a public body which will be beyond suspicion."[41] The editor of *Scientific American* expressed his skepticism to one of the study's authors:

> I may say that the results which you are getting seem to me very surprising. Analysis of ethyl gas, made independently by our chemical experts, indicate[s] that the treated gasoline contains approximately 2 grams of metallic lead per gallon of gasoline. This lead has to come out somewhere, and one would expect

it to come out in the form of either powdered carbonate or powdered oxide. Both of these powders are well known to be extremely poisonous. Therefore, it seems to me very remarkable that you have had no symptoms of poisoning in your exposed animals with the amount of exposure to the exhaust gases which you have used.[42]

The Ethyl Corporation pushed back, bringing out their own experts to corroborate the study's findings. The industrial hygienist Dr. Emery Hayhurst was on the editorial committee of the esteemed *American Journal of Public Health*, and published an editorial in that journal stating that the evidence supported "'complete safety' so far as the public health has been concerned." Readers of the journal would have had no way of knowing that Dr. Hayhurst was working as a consultant for the Ethyl Corporation. In fact, Dr. Hayhurst had also worked as an adviser to the Workers' Health Bureau, and private correspondence has since shown that he was secretly handing over information about their plans and strategies to federal employees. Basically, he was a spy.[43]

Six months after the release of the Bureau of Mines study, in May 1925, the surgeon general convened a conference on TEL in Washington, D.C. The eighty-seven participants included public health experts and labor advocates, as well as representatives of the oil and automobile industries. These groups came to the conference with very different perspectives on the safety of TEL and on the appropriate standards to use in deciding whether it should be banned.[44] The health and labor side attacked the methodology of the Bureau of Mines study, and brought their own data suggesting that the manufacture and use of TEL posed serious hazards to the workers making it, as well as to everybody who would be exposed to the lead coming out of America's tailpipes. The physiologist Dr. Yandell Henderson presciently warned that "conditions would grow worse so gradually and the development of lead poisoning will come on so insidiously . . . that leaded gasoline will be in nearly universal use and large numbers of cars will have been sold . . . before the public and the government awaken to the situation."[45]

Occupational health expert Dr. Alice Hamilton also tried to direct the attention of the conference toward the potential long-term effects of lead accumulation on the general public. "You may control conditions within a factory," she said, "but how are you going to control the whole country?"[46] Unfortunately, the anti-lead side struggled to be taken seriously, partly because of issues of social status. The men killed at TEL plants had included a radical Irish immigrant and a Black janitor, and the experts who gathered to argue on their behalf included several women and an outspoken pacifist. Dr. Hamilton herself worried that the anti-lead argument that she and her colleagues put forth was undermined by "an unmistakable aura of Socialism or feminine sentimentality for the poor."[47]

On the other side of the debate were titans of industry, who claimed that TEL was an important and irreplaceable product that should not be banned because of unwarranted fears. Charles Kettering and others claimed during the conference that no alternatives were available, though historian William Kovarik has demonstrated that they knew this statement to be false.[48] In addition, they argued that harm to workers could be avoided with proper safety measures, and that there was no risk of harm to the general public. One conference participant, Frank Howard of Standard Oil, had this to say about the issue:

> We cannot quite act on a remote probability. We are engaged in the General Motors Corporation in the manufacture of automobiles, and in the Standard Oil Company in the manufacture and refining of oil. On these things our present industrial civilization is supposed to depend. I might refer to the comment made at the end of the war—that the Allies floated to victory on a sea of oil–which is probably true. . . .
>
> Now . . . we have this apparent gift of God—of 3 cubic centimeters of tetraethyl lead—which will permit that gallon of gasoline . . . to go perhaps 50 percent further. . . .
>
> What is our duty under the circumstances? Should we throw this thing aside? Should we say, "No, we will not use it," in spite of the efforts of the government and the General Motors Corporation and the Standard Oil Co. toward developing this very thing, which is a certain means of saving petroleum? Because some animals die and some do not die in some experiments, shall we give this thing up entirely?
>
> Frankly, it is a problem that we do not know how to meet. We cannot justify ourselves in our consciences if we abandon the thing. I think it would be an unheard-of blunder if we should abandon a thing of this kind merely because of our fears.[49]

Howard's testimony speaks volumes. One of the most common questions I've been asked since I started working on this book is, "If we knew that lead was bad for people, why did we put it in gasoline in the first place?" It's true that the product that the General Motors and Standard Oil executives were pushing would go on to do incalculable harm to generations of American children. That is exactly why it is so important to understand the belief systems and economic incentives that influenced these "upstanding" businessmen to make the arguments that they did in favor of leaded gasoline, and perhaps even to convince themselves that they were doing the right thing. (Frank Howard would later go on to collaborate illegally in wartime with Nazi Germany,[50] but most of the industry representatives at the 1925 surgeon general's conference weren't actually Nazi sympathizers, just hard-nosed industrialists.)

Economic incentives would have nudged automobile manufacturers and oil company executives toward dismissing any potential harms from leaded gas. It is easy to assume that only the morally corrupt are swayed by economic incentives, but the research suggests that the influence of economic incentives runs deep throughout mainstream American culture. For example, a number of studies have shown that people are significantly more likely to donate blood—which saves lives—if they are rewarded monetarily for doing so.[51] Other studies show that people are much less likely to litter if they can get a nickel for returning their bottle or can instead,[52] and that charitable giving is strongly influenced by tax incentives.[53] As much as we may want to believe otherwise, our moral convictions are often influenced by economics.

Everyone involved in the production and use of leaded gasoline stood to make a lot of money if the product was allowed to be sold. TEL raised the octane of gasoline very cheaply, and the Ethyl Corporation held the exclusive patent. Frank Howard, Charles Kettering, and Thomas Midgley all knew that ethanol represented a viable alternative—Midgley himself had done extensive work on alcohol/petroleum mixtures—but none of them were going to get rich selling ethanol additives.[54] Upton Sinclair famously asserted that "it is difficult to get a man to understand something when his salary depends upon his not understanding it."[55] The industry executives at that 1925 conference were running companies that provided salaries to thousands of workers and their families, as well as the substantial salaries of the executives themselves. So they falsely claimed that TEL, and only TEL, could provide the octane boost that the United States needed to run its increasingly petroleum-dependent economy.

In addition, Howard's testimony highlights the argument, made by multiple representatives of the oil and automobile companies, that all new technologies have some risks, and for civilization to move forward, we just have to accept those risks. By raising issues of patriotism and national defense, Howard's speech even suggests a context in which facing the perils of this bold new undertaking might be seen as brave and noble. Workplace hazards were common at the time, and workers were expected to accept some dangers along with their paychecks. One industry expert admitted that the manufacture of TEL had harmed workers, but argued that "its casualties were negligible compared to human sacrifice in the development of many other industrial enterprises."[56] This was an era when reformers were pushing for better workplace health and safety protections in a variety of industries, and those industries were pushing back hard.[57]

It is also unsurprising that the businessmen at the surgeon general's conference would try to keep the focus on the acute effects of lead exposure, not the chronic ones. Acute effects are the sudden, severe symptoms associated with a large, short-term dose of lead, such as those experienced by the workers who died or went crazy in poorly ventilated TEL factories. Chronic effects are the subtler symptoms that are a result of long-term exposure to lower levels of lead. Proponents of TEL argued that the problem of acute lead poisoning could be solved through better safety training for their workers.[58] They had no such easy response to the potential for chronic effects among the population at large.

Their willingness to accept a significant level of risk to their employees, as well as their exclusive focus on acute effects, means that the industry representatives at the conference weren't even asking the same question as the public health experts. The oil and automobile manufacturers, through the Bureau of Mines study and their own in-house research, had set out to answer the question "Can we produce TEL without our workers suffering from the severe neurological and gastrointestinal symptoms associated with acute lead poisoning?" They determined that the answer was "probably," and considered that good enough to go ahead. Public health experts were interested in the question "Will there be long-term harm to workers and the general public from the widespread manufacture and use of leaded gasoline?" Since the federal Public Health Service had declined to research the issue, there wasn't a lot of information available to answer that question. Frank Howard's speech makes it clear that, even if the answer to this second question was "possibly," the producers of leaded gasoline didn't think that was a good enough reason to stop manufacturing this "gift of God."

The difference in perspective between the two sides at the surgeon general's conference in 1925 was primarily one of priorities. In the face of evidence that was, at the time, uncertain and incomplete, the public health experts put a priority on protecting the public from potential harms, while the company executives put a priority on economic gains. This led one side of the debate to argue that a product should not be used until proven safe, and the other side to argue that a product should not be banned unless proven unsafe,[59] a debate that has continued to dominate the discussion of environmental health issues throughout the past century.

At the end of the 1925 conference, participants on both sides of the debate supported a resolution for the surgeon general to form a special committee, consisting of six medical experts from top universities, to study the issue. The committee was given only a few months to reach a conclusion, so they were forced to rely on the Ethyl Corporation itself for some of their data. For example, based on results from a flawed engine test, one of the committee members made a "very tentative" estimate about how much lead would end up in the air, an estimate that would eventually prove to be more than one hundred times too low.[60] After a mere seven months, the committee released a report that found "no good grounds" for prohibiting the sale of leaded gasoline. In reporting the committee's findings, the *New York Times* ran the headline "Report: No Danger in Ethyl Gasoline." However, that's not exactly what the committee's report said. In fact, the report specifically noted that "it remains possible that if the use of leaded gasolines becomes widespread, conditions may arise very different from those studied by us which would render its use more of a hazard than would appear to be the case from this investigation. Longer experience may show that even such slight storage of lead as was observed in these studies may lead eventually in susceptible individuals to recognizable or to chronic degenerative diseases of a less obvious character."[61]

Along with issuing their report, the special committee passed a resolution calling on the Public Health Service to conduct further studies, arguing that "it should be possible to follow closely the outcome of a more extended use of this fuel and to determine whether or not it may constitute a menace to the health of the general public after prolonged use or other conditions not now foreseen."[62] However, the PHS ignored this recommendation, and did not continue the work begun by the special committee. In fact, no serious government research on the health effects of leaded gasoline would be carried out for the next four decades.

And just like that, Standard Oil gas stations all over the country put up signs proclaiming that "Ethyl is Back." In one small concession to worker safety, the TEL was now being mixed into gasoline at the refineries rather than at the pump.[63] American cars became bigger, faster, and more powerful, and used more and more leaded gasoline. By 1936, around 90 percent of gasoline sold nationwide was leaded. The amount of lead that Americans were putting into their gas tanks would climb steadily across the decades until, by the early 1970s, over 250,000 tons of lead per year were being added to gasoline in the United States.[64] Most of the lead that's pumped into gas tanks eventually comes out of tailpipes, in the form of airborne lead-containing chemicals. The air and soil throughout the country would come to contain higher and higher concentrations of lead over the decades. And so would the children.

3

Getting the Lead Out

▬ ▬ ▬ ▬ ▬ ▬ ▬ ▬ ▬

During the decades that the United States was burning leaded gasoline, just about every tailpipe in America was releasing tiny, invisible particles of lead oxides. All that lead floated through the air, and what wasn't breathed into somebody's lungs eventually settled out on a surface, creating lead dust. When that dust got onto children's hands, as they played outside or when it was tracked into the house, some of it inevitably ended up in their mouths. Through both inhalation and ingestion, kids were taking in a lot of lead.

Eventually scientists, activists, and policy makers would embark on a contentious debate about all that lead, and what to do about it. The decades-long fight to regulate leaded gasoline—as well as other sources of lead exposure— included arguments about how much lead was actually getting into people's bodies, and about what levels of lead might be considered "natural," or deemed "safe." So let's take a moment to look at how lead in the body is measured, and what those measurements mean.

Lead can travel throughout the body. It can get into different organs, including the brain, and it builds up in the bones and teeth. The most straightforward way to measure the amount of lead that a person has in their system at a particular time is to look at the level of lead in their blood. Blood lead level is typically measured in micrograms (μg) per deciliter (dL). A deciliter is pretty easy to understand—it's a tenth of a liter. So if you have a one-liter bottle of water or soda, just imagine dividing it into ten small cups. Or if tequila is your beverage of choice, think about two standard shots—one for you and one for a friend, and that's a deciliter.

Micrograms, on the other hand, are hard for us to really get a good sense of. A whole gram is already pretty small—about the weight of a single paperclip, or half a dime. A milligram is one-thousandth of that, somewhere around the weight of a single grain of salt. The unit we use to measure blood lead level— the microgram—is only one-thousandth of a milligram. It would take a thousand of them to make up a single grain of table salt. So, whether a measurement is five micrograms or twenty micrograms in each tenth of a liter of blood, it may seem like a really small amount.

Still, when we're talking about powerful chemicals like lead, every microgram packs a punch. Two micrograms of LSD in every deciliter blood, for example, is enough to have seriously mind-altering effects on a full-sized human. I, personally, suffer from hypothyroidism, a disorder caused by my thyroid gland not making enough of certain hormones, leading to fatigue and weight gain, among other symptoms. Luckily, my hypothyroidism is easily treated with a small daily pill that contains about one single microgram of synthetic hormones for every deciliter of my blood. A few micrograms can make a big difference, and we now know that a few micrograms of lead per deciliter of blood (written as "µg/dL") can make a huge difference.

Early in the twentieth century, our techniques for measuring lead in blood weren't very precise. As the techniques improved and company doctors started measuring the blood lead levels of potentially lead-poisoned workers, it was common to use a threshold of 80µg/dL or even 100µg/dL to define lead poisoning. At these levels, lead can cause severe constipation, anorexia, convulsions, inability to control one's limbs, paranoid delusions, coma, and even death.[1] Eventually, it would become clear that lead is harmful at much lower levels, especially to children's developing brains. In the 1950s and 1960s, different screening programs used 80, 60, or 50µg/dL as their cutoff. In the 1970s, the federal government set its level of concern at 30µg/dL, a value that would later be reduced to 25µg/dL, then 10µg/dL.

Finally, in 2012, the Centers for Disease Control (CDC) began using a "reference level" of 5µg/dL. When a child's blood lead level is at or above this reference level, the CDC recommends "environmental investigations" to find the source of lead exposure, along with follow-up testing.[2] It has now been widely demonstrated that levels below 5µg/dL can still be harmful to children, and some have argued that pediatricians should intervene when children are found to have blood lead levels of 2µg/dL or higher.[3] In fact, a "consensus statement" endorsed by fifteen national or international medical organizations found that "scientists agree that there is no safe level of lead exposure for fetal or early childhood development,"[4] and a May 2019 CDC report states that "no safe level has been identified."[5] As we'll see in chapter 5, even a tiny bit of lead in the brain can harm the delicate process of neurological development.

As we look back at the debate over lead exposure and lead regulations, it's useful to keep our current understanding of blood lead levels in mind. When my daughter turned one, she had her blood lead level tested for the first time. If her result had come back above 1μg/dL, I would have been concerned, because there's a lot of good evidence that even 2μg/dL is harmful to children. (How, exactly, does lead harm children? We'll look at the specific health effects of lead in chapter 5.) If her blood lead level had come back at 5μg/dL or more, her pediatrician would have helped us to identify potential sources of exposure, which we would have worked diligently to eliminate, and she would have been tested again in a few months. If it came back at 20μg/dL or higher, she'd have been taken to the hospital for further testing, including X-rays, more blood work, and a neurological assessment. (Luckily it came back at 0μg/dL, which means her blood lead level was below the amount that's detectable by the test. Whew!) We now know that there is *no safe level* of lead exposure. By today's standards, a blood lead level of 2μg/dL is potentially harmful, a level of 5μg/dL or more requires intervention, and anything over 20μg/dL is an urgent problem.

Knowing what we know now, it's easy to be shocked by the idea that a blood lead level of 79μg/dL was ever considered "safe," but for a long time many people thought it was. Some early lead research was interpreted as suggesting that anything below 80μg/dL was harmless. In the first half of the twentieth century, public funding for scientific research was scarce, so most science was privately funded. Throughout the 1930s, 1940s, and 1950s, almost all of the research on lead exposure and its health effects came from one place—the lead industry itself. In fact, a surprising amount of that research was done by one single person—Dr. Robert Kehoe.[6] Dr. Kehoe was employed for most of his career by the very company that produced lead for gasoline, and we have to assume that his contention that lead causes no harm below a "threshold" of 80μg/dL was influenced by the fact that he worked for the lead industry.

In 1923, Dr. Kehoe was only three years out of medical school, doing research and teaching at the University of Cincinnati. His research involved studying the effects of various toxins—including lead—on proteins. The head of the physiology lab where he worked was an old friend of Charles Kettering's, so when Kettering and Thomas Midgley went looking for a medical expert to investigate the spate of lead poisonings among the workers at their tetraethyl lead (TEL) plants, they hired the young Dr. Kehoe. By 1925, Kehoe was named medical director of the Ethyl Gasoline Corporation, a position he would hold for more than thirty years, even as he continued to serve as a faculty member at the University of Cincinnati.[7]

Dr. Kehoe's early work for Ethyl probably saved a lot of lives. He developed safety procedures for every step of the TEL manufacturing and distribution process, mandating improved ventilation, protective equipment for the workers, and strict control over each gallon of TEL that Ethyl produced and shipped.

Workers in TEL plants became much less likely to suffer from acute lead poisoning. It's easy to see how his early successes protecting workers from untimely death could have influenced Kehoe's views about lead and lead poisoning. From his perspective, lead was mainly an occupational hazard, and the primary goal was to prevent the serious consequences of acute lead poisoning—the cramps and nausea, convulsions, and hallucinations of the severely poisoned worker. Unfortunately, this focus on adult workers and acute effects would lead Kehoe to draw some erroneous conclusions that hindered the progress of research on lead exposure for decades.[8]

In the 1930s, Kehoe did a series of studies in which he fed different amounts of lead to healthy volunteers, and then measured the lead in their blood, urine, and feces. (As a member of my university's Institutional Review Board, I find it appalling that the standards of that time allowed Kehoe to intentionally expose healthy research subjects to a known toxin just to see how they metabolized it. We require our researchers to implement a complicated process of informed consent just to have their subjects complete a pencil-and-paper survey, but that kind of oversight of research wasn't required until the 1970s.[9]) He calculated that the research subjects excreted nearly as much lead as they took in, maintaining an "essential balance." Kehoe's interpretation of the results of these studies was that the human body has protective mechanisms that allow it to get rid of any excess lead that it takes in. We now know that these studies were rife with measurement inaccuracies and contamination, but Kehoe didn't know that at the time.[10]

In addition, Dr. Kehoe measured the amount of lead in the bodies of many different people and found that everybody he tested contained a substantial amount of lead. He especially highlighted his work with "inhabitants of two isolated Mexican villages, far removed from any possible contact with lead, except through the use of earthenware utensils glazed with a lead compound." It seems absurd that Kehoe considered these rural Mexican research subjects—whom he offensively contrasted with "normal civilized adults"—to represent a baseline level of lead, ignoring the fact that they were regularly eating off of lead-glazed pottery, but that was, in fact, his interpretation. In comparison with these isolated Mexican villagers, he came to the following conclusion about the subjects he had tested in the United States: "The occurrence of lead in food and in human excreta, in the amounts found, seems to be neither a new nor a localized phenomenon, and it presents no new hygienic problem."[11] From these studies, Kehoe concluded that lead in the body is normal and natural, and nothing to be concerned about.

Kehoe's research on workers in TEL factories led him to conclude that there is a "threshold" level of lead, below which no lead poisoning occurs. Because he did not see patients with clear symptoms of acute lead poisoning whose blood lead levels were below 80μg/dL, Kehoe concluded that this was a "safe" level of lead in the body. In fact, he argued that levels well above 80μg/dL may be safe for some workers, because he had seen higher numbers in some workers who still showed no overt symptoms of lead poisoning. As late as 1965, Kehoe was still arguing that lead poisoning does not occur at blood lead levels lower than 80μg/dL.[12]

This idea of a threshold below which lead is perfectly safe would prove to be an enduring one. Even when later researchers argued that the level should be lower, especially for children, many of them still subscribed to the belief that there is a definitely "safe" level of lead. If a substantial exposure to lead is a long-standing part of the normal human environment, then it makes sense that our bodies would have developed mechanisms for metabolizing and excreting any excess lead. If we can effectively process and expel excess lead, then it makes sense that low-level exposures would be no problem for our bodies to deal with. There would have to be some threshold above which our natural lead removal process breaks down, but anything below that level should be essentially harmless. Kehoe's theory of lead exposure was clear and compelling, and seemed to be supported by the evidence. Case closed.

Only, he was wrong. There is no safe level of lead, because our bodies aren't adapted to deal with any substantial amount of lead in our environment. We have no evolutionary defenses against lead because it was hardly present at all in the preindustrial environment. Kehoe and his fellow researchers didn't know that at the time. When they went to look for evidence of lead far away from modern industrial sources, they found it—lots of it—even when they weren't studying populations whose pottery was lead-glazed. Unfortunately, their work was undermined by a problem that is the curse of laboratory scientists everywhere, from Chemistry 101 students to Nobel Prize winners—contamination. Of course they found lead everywhere they looked; they themselves were covered in the stuff. Their laboratories, their equipment, even their hair and skin, were coated with those lead oxide particles that were pouring out of the nation's tailpipes and into its air. They thought that they were looking at a natural environment that's full of lead, but they actually couldn't see past their own mess.[13]

For more than three decades, Kehoe would remain supremely confident in his own erroneous model of lead poisoning. Because his laboratory was the only place where considerable research on lead exposure was being done for much of that time, his word carried a lot of weight. In 1966, he claimed in a Senate committee hearing that he and his colleagues "had been looking for 30 years for evidence of bad effects from leaded gasoline in the general population and had found none." Even more strikingly, he boasted at that hearing that he knew more about lead than anybody else in the world, and asserted that "at present, this laboratory is the only source of new information on this subject." By the mid-1960s, this was no longer quite true, but by that time, many people considered Kehoe to be *the* authority on lead exposure, and were willing to take his word for it.[14]

The fact that one man, running one laboratory, could dominate our understanding of lead poisoning—and therefore our policies about lead—for so long highlights the logical fallacy of argument from authority. It's true that the credibility of a source is important to take into consideration when evaluating evidence, and Kehoe did seem to be a supremely credible source—he held a prestigious university position and was the author of many papers published in well-regarded scientific journals. However, it's also important to evaluate the evidence itself, and unfortunately a lot of people were inclined to believe Dr. Kehoe long after his work had been undermined by new evidence.

The person most responsible for undermining Dr. Kehoe's work was the inimitable Dr. Clair Patterson. Dr. Patterson was a geochemist who didn't set out to study lead poisoning at all; he set out to determine the age of the Earth. During World War II, Patterson had worked on the Manhattan Project in Oak Ridge, Tennessee, where he learned a brand new technique for measuring tiny amounts of different types of uranium. After the war, he got to work using this new technique to measure minute quantities of uranium and lead in rocks and meteorites. Scientists had discovered that the ratio of a certain uranium isotope to a certain lead isotope acts like a clock—it shifts over time in a predictable way. So, if you can measure that ratio, you can determine how old a rock is. The tricky thing is that you have to be very, very precise. Patterson kept finding too much lead, and he wanted to know why.[15]

Once he started looking for sources of lead contamination, Patterson found them everywhere. In order to get his samples contamination-free enough to make his measurements, he had to become a fanatic about cleaning. In fact, Patterson is widely credited with inventing many of the clean laboratory protocols that are in common use today. It took him seven years, but he finally produced the clean samples he needed, and became the first person to accurately calculate the age of the Earth as 4.5 billion years. He was so excited by his discovery that he thought he was having a heart attack, and asked his mother to take him to the hospital.[16]

Before long, Dr. Patterson turned his attention to measuring and characterizing the very lead contamination that had so vexed his age-of-the-Earth project. The clean lab techniques he had developed allowed him to produce much more accurate measurements of lead than other labs had ever done. He found that the surface layer of the ocean contained substantially more lead than deeper layers did, demonstrating that more and more lead was entering the ocean every year. These results were published in 1963, in a paper that argued that the excess lead came from leaded gasoline fumes. That was the first evidence of just how widespread the contamination from leaded gas really was.[17]

Patterson's work on lead would take him all the way to the ends of the earth. In the mid-1960s, Patterson led a team that collected ice core samples from Greenland and Antarctica and used them to measure changes in lead levels over time. He also measured lead in rocks, soils, plants, animals, air, and water. He compared the lead levels in modern Americans with the amount of lead in the bones of ancient Peruvians. Again and again, his results demonstrated that the amount of lead in our environment, and in our bodies, was far, far more than the natural background level. In 1965, Dr. Patterson declared, in an article in *Archives of Environmental Health*, "The average resident of the United States is being subjected to severe chronic lead insult."[18]

In using such strong and decisive language, Dr. Patterson had made himself a target. Representatives of the lead industry worked hard to undermine not just Patterson's results and his interpretations, but Patterson himself. One Ethyl executive called Patterson a "crackpot," and discussed how to challenge him in an internal company memo: "Patterson is playing fast and loose with chemistry and physiology. Since he is an authority in neither field, he should be vulnerable in both."[19] Others called Patterson a "nut" and a "kook." This dismissal of Patterson's work demonstrates the danger of ad hominem attacks, of attacking the messenger instead of the message. Patterson was, indeed, eccentric. Gangly and awkward, he never came across as smooth as the industry "experts" who were belittling his work. He was known as a long-winded and unclear speaker. Even some of his biggest supporters have described him as bullheaded and difficult to get along with. His behavior often bucked social norms—for example, he was known to wear a gas mask while walking around the Caltech campus on smoggy days. Perhaps Patterson was a bit of a "kook." But history would show that he was right about lead.[20]

Patterson's work undermined one pillar of Kehoe's theory—the idea that the amount of lead released by cars burning leaded gas isn't very different from the natural background level. If ancient humans weren't historically exposed to much lead, that raises questions about Kehoe's assertion that we have evolved metabolic processes that protect us from any harmful buildup of the element. Meanwhile, other researchers were starting to undermine Kehoe's idea that blood lead levels below 80µg/dL are "safe." Some of the most important early research on low-level lead exposure was done by a gifted pediatrician named Dr. Herbert Needleman.

Early in his career, in the 1950s, Dr. Needleman's work at the Children's Hospital of Philadelphia included treating lead-poisoned children from nearby poor neighborhoods. Needleman was dismayed to learn that the hospital sent these kids back home to the same apartments, with their crumbling plaster and peeling paint, where they had gotten lead-poisoned in the first place. Later, after he trained to be a psychiatrist, Needleman worked with families in some of these same poor neighborhoods. He saw children with behavior problems and difficulties in school, and he kept wondering—could these kids be suffering from the effects of undiagnosed lead poisoning? He wanted to find out.[21] (It's possible that Dr. Needleman had been primed to look for the effects of lead poisoning by his work one summer during medical school as a laborer at one of the DuPont plants where lead workers had died back in the 1920s, and where lead-poisoned workers had been pointed out to him by colleagues.[22])

Dr. Needleman knew that measuring lead in blood only tells you about children's current exposure, not their overall lifetime exposure to lead. Lead is stored in the bones, so he considered trying to examine lifetime lead exposure through bone biopsies, but these are invasive and extremely painful. Hair and fingernails are too contaminated by outside sources to accurately measure the lead inside them. Finally, he hit on the perfect way of estimating lifetime exposure to lead in children—baby teeth. Needleman and his colleagues collected baby teeth from hundreds of kids in neighborhoods around Boston. (In the scientific literature, they're called "deciduous" teeth, like leaves that drop in the fall.) The researchers even provided a special note that the kids could put under their pillow to inform the Tooth Fairy that they had donated a tooth to science, so they wouldn't miss out on the reward. Each tooth was carefully cleaned, a tiny fragment was removed, and the amount of lead in that fragment was measured using an electrochemical technique.[23]

The results of Dr. Needleman's tooth study, published in the prestigious *New England Journal of Medicine*, were striking. The kids in his "high lead" group performed significantly less well on measures of intelligence and attention than the kids in his "low lead" group, even when the researchers controlled for potentially confounding factors like parental IQ and income level. Teachers also rated the "high lead" kids as having more classroom behavior problems than the "low lead" kids. (The teachers weren't told the lead results so that their ratings wouldn't be biased by them.) Because the study had excluded any kids with diagnosed lead poisoning, these were all kids whose blood lead levels were below the level that industry experts were arguing was "safe." Needleman's later work would follow another group of kids over time, demonstrating that the effects of lead exposure in early childhood are still measurable well into the teenage, and even adult, years.[24]

Like Dr. Patterson, Dr. Needleman was the subject of some harsh ad hominem attacks. Certain other scientists disagreed with the statistical techniques he had used to analyze the data in his baby tooth studies. That's a normal part of scientific discourse—arguing over the proper methods for analyzing complex data. In this case, scientists who disagreed with Needleman, some of them on the payroll of a lead industry trade association, went much farther. They accused him of academic misconduct and tried to get him fired from his university position.[25] Eventually, Needleman was cleared of all wrongdoing, but it must have been very stressful for him and his family to have his livelihood and integrity threatened like that.

Dr. Needleman wasn't the only researcher discovering new dangers associated with lead. In experimental studies on animals and epidemiological studies of human populations, many scientists were demonstrating that lead causes real and lasting damage to the nervous system at much lower levels than previously thought. The final pillar of Robert Kehoe's theory could no longer stand. Researchers had clearly demonstrated harm at levels well below Kehoe's "safe" threshold of 80µg/dL. The cutoff for considering kids to be lead-poisoned was lowered, and lowered again. Finally, some scientists began to argue against the very existence of a threshold, concluding that there was no level of lead in children's bodies that could be considered "safe."

During the same time as this dramatic shift in society's understanding of lead exposure, there was a transformation in the national conversation about environmental toxins more generally, and the role of the federal government in protecting us from those toxins. Back in the 1920s, when lead was first put in gasoline, there was no federal agency tasked with protecting us from toxins in the environment. Many people felt that the role of the government was to ensure fair business practices in a competitive marketplace, and otherwise not to infringe on the activities of corporations. By the 1960s, though, the times, they were a-changin'.

The modern environmental movement really took off with the publication of *Silent Spring* by Rachel Carson in 1962. Carson captured national attention with her descriptions of the harm that widespread spraying of DDT was doing to our environment, and potentially to our bodies as well. Her work was part of a groundswell of interest in the 1950s and 1960s in protecting the environment and human health, and Americans were coming to see a role for the government, at local, state, and federal levels, in enforcing health and environmental protections.

The Clean Air Act was passed in 1963, and the federal Environmental Protection Agency was created in 1969. Suddenly, the federal government had a greatly expanded role in protecting Americans from dangerous chemicals, products, and workplaces. Between 1968 and 1978, Congress passed more regulatory statutes than they had in the country's entire previous 179 years.[26] These regulations had hugely beneficial impacts on the lives of all Americans. The air, the water, and the food supply got cleaner. Homes, factories, and cars got safer. Finally, the regulatory apparatus was in place to tackle the problem of lead poisoning.

As the need for regulations to protect children from lead poisoning became apparent, the Ethyl Corporation and its allies argued that lead paint was the sole culprit, so there was no need to regulate gasoline. Many researchers at the time believed that the lead in the dust and soil that kids were coming into contact with came almost exclusively from disintegrating lead paint. In Baltimore, Dr. Howard Mielke carried out an experiment in the 1970s designed to assess the spread of lead paint particles into the surrounding environment. He measured dust and dirt in parks and yards throughout Baltimore and the surrounding suburbs. Nearly all the buildings in the center of the city had been rebuilt out of unpainted brick way back in 1904, after a major fire, so Dr. Mielke expected to find lower levels of lead in the city and higher levels out in the suburbs, where most of the buildings were painted wooden houses. Instead, he found the exact opposite.[27]

Dr. Mielke told me about this study when I went to visit his current lab at the Tulane University School of Medicine in New Orleans. (We'll hear more about Dr. Mielke's recent research in New Orleans in chapter 6.) At first, he was stumped by his own results. Why would there be so much more lead on the ground in the inner city, where the brick buildings weren't shedding any paint dust? Dr. Mielke told me about his "aha" moment: "I was driving around and looked at the gas pump, and it said, 'contains lead.'" So Dr. Mielke compiled data on how much gasoline was sold each year in the state of Maryland, then multiplied that number by the amount of lead per gallon, and demonstrated that the extra lead contaminating the yards and playgrounds in the inner city was coming out of the tailpipes of Baltimore city traffic.[28]

Dr. Mielke told me that he had difficulty in getting these results accepted for publication in a scientific journal. Typically, journals send prospective articles out to other experts in the field, who review them anonymously and send in critiques that need to be addressed before the paper can be published. Dr. Mielke's paper kept coming back with odd comments from reviewers, and he believes that these were industry-funded scientists who were attempting to keep the paper from getting out. When the paper was finally published, critics argued that Baltimore is a unique case because of all the heavy industry located there. However, Dr. Mielke was able to replicate his results in Minneapolis, which was relatively free of heavy industry—he found that there was still more lead downtown, where the traffic was. (During our interview, I was startled to realize that, during the years in the 1970s when Dr. Mielke was finding hazardous levels of lead in the soil in parks and yards around Minneapolis, I was a toddler crawling around in the parks and backyards of that same city.[29])

Over a delicious seafood dinner at his favorite local restaurant in New Orleans, Dr. Mielke told me the story of his experience when he was asked to testify before Congress about the threat from leaded gasoline. In the "crowded and chaotic" hearing room, he had trouble getting to the press table to hand over his written materials so that his testimony and supporting evidence could be covered accurately in the news media. Somebody came up and offered assistance, so Dr. Mielke handed over his written materials to the helpful stranger. That stranger turned out to be a lawyer for the Ethyl Corporation who, instead of delivering the papers to the press table as promised, confiscated them and left the hearing room. The Ethyl Corporation and its allies were not keen on letting the public hear from a researcher who was determined to undermine their contention that lead paint was the only important source of childhood lead poisoning.[30]

By this time, some progress had already been made in addressing the problem of lead paint. The amount of lead used in paint had peaked in the 1920s, and had been falling ever since. The first U.S. regulation addressing lead paint was a ban on its use indoors in Baltimore in 1951. Soon after that, in 1954, the paint industry set a voluntary standard that indoor paint should contain no more than 1 percent lead when dry. Remember that some early twentieth-century paints had been as much as 50 percent lead.[31] National legislation finally came along in 1971, with the Lead-Based Paint Poisoning Prevention Act, which officially banned any paint containing more than 1 percent lead for any housing constructed or renovated with federal assistance. An amendment to this law reduced the acceptable level of lead to 0.06 percent in 1974, and in 1978 the Consumer Product Safety Commission extended this limit to all paint sold for residential use.[32]

Thus, by the late 1970s, the problem of *new* lead paint had been nearly eliminated. The challenge was all the old lead paint that was still in place, and still poisoning children every day. This is a serious, ongoing problem, and we'll talk more about it in chapter 7. However, across the country, the average amount of lead on the walls of American children's homes has been declining for almost a century. We still have a long way to go, but we've been moving in the right direction for a long time. Similarly, existing lead pipes still pose an ongoing threat (as we'll see in chapter 7), but the amount of lead in pipes and plumbing fixtures declined throughout much of the twentieth century. Lead solder for food cans wasn't banned until the 1990s, but by that time its use had also been declining for many years. Yet, as the modern environmental movement coalesced in the 1960s, there was one source of childhood lead exposure that was still skyrocketing—leaded gasoline.

Way back in 1926, when the Office of the Surgeon General approved the sale of leaded gasoline, the Ethyl Corporation had agreed to a limit of 3mL of tetraethyl lead (TEL) per gallon. (That's a little over half a teaspoon.) But cars kept getting bigger and more powerful, so in 1958, Ethyl asked for approval to raise that amount to 4mL per gallon. A new surgeon general's committee approved this request, but as part of its conditions, it formed a Working Group on Lead Contamination, which carried out the first major study on the effects of leaded gasoline in over three decades.[33] By that time, leaded gasoline was everywhere—over 98 percent of the gas sold in the United States was leaded, and throughout the 1960s, the total amount of lead we put in our gasoline kept going up every year.[34]

At first glance, the Working Group's study seemed to favor the leaded gasoline industry. Although they found more lead in cities than in rural areas, their sampling found lead everywhere in the environment, which Robert Kehoe and others argued was because lead is "natural." Not a single subject in the study had a blood lead level above the 80μg/dL standard that the industry supported, allowing them to claim that leaded gasoline was not a health risk. However, a number of subjects were found to have blood lead levels above 60μg/dL, which worried many scientists, including members of the Working Group.[35] After this study, the 4mL/gallon standard was allowed to continue. The Ethyl Corporation had won the battle, but they may have inadvertently started the war that would eventually lead to the demise of their product.

In 1966, a congressional clean air subcommittee held hearings on the issue of leaded gasoline, and Robert Kehoe was there, once again presenting himself as *the* expert on the issue: "It so happens that I have more experience in this field than anybody else alive."[36] He made his usual claims that lead is natural, the body eliminates it efficiently, and any lead level below 80μg/dL is basically harmless. This time, Clair Patterson was there too, presenting his recently published work showing that the amount of lead in the modern environment was far from "natural," and contending that it represented a serious health threat. The chairman of the subcommittee, Senator Edmund Muskie, asked Patterson why researchers like Kehoe had not attempted to distinguish between "typical" and "natural" levels of lead. Muskie pointed out that making this distinction "seems like such a logical approach." Patterson replied, "Not if your purpose is to sell lead."[37]

It wasn't just lead; many forms of pollution that had been previously unregulated were coming under scrutiny during the 1960s. The first serious legislative blow to leaded gasoline would actually come from another issue entirely—the problem of smog. Smog had become a major concern in American cities, reducing visibility and contributing to respiratory disease and other illnesses. With the passage of the Clean Air Act in 1963, the federal government began regulating air quality for the first time. Later amendments to the act zeroed in on cars and trucks as a major contributor to smog, and set increasingly stringent limits on the amount of smog-causing chemicals that could come out of America's tailpipes. In order to meet these new standards, the automotive industry developed a device called the catalytic converter.[38]

A catalyst, as you probably learned in a high school chemistry class, is a substance that speeds up a chemical reaction. The chemical reactions in this case destroy some of the toxic, smog-producing chemicals released by internal combustion engines. The catalyst in a catalytic converter is platinum, the same metal in the record they give you if you sell a million albums. Here's the problem—lead actually destroys this platinum catalyst. As long as cars were burning leaded gasoline, they couldn't use this new technology to reduce smog.

For years, automobile manufacturers fought alongside the Ethyl Corporation against any requirement to install catalytic converters, but eventually some of them bowed to rising public pressure and embraced the technology. In 1972, the Environmental Protection Agency (EPA) announced that it would mandate catalytic converters on all new cars starting in 1975, and all those cars would need unleaded gasoline to keep the platinum catalyst in their catalytic converters working. So the EPA also required that every gas station in America would sell at least one grade of unleaded gasoline from 1975 forward.[39] The EPA even mandated special nozzles for unleaded gasoline, so that people wouldn't accidentally put leaded gas in their new cars and ruin their catalytic converters.[40]

Of course, there would still be millions of pre-1975 cars on the road, burning leaded gas. By this time, experts like Dr. Patterson and Dr. Needleman had made a compelling case about the need to reduce lead exposure to protect the health of all Americans, especially children. So the EPA proposed another regulation in 1972, mandating a major reduction in the lead content of the entire gasoline pool. However, in the face of industry pushback, the EPA dragged its feet on setting a specific schedule for the drawdown. So the Natural Resources Defense Council (NRDC), a nonprofit environmental advocacy group, sued the EPA in federal court. The NRDC won, and in December 1973, the EPA mandated reducing the amount of lead in all gasoline.[41] Between 1972 and 1980, the total amount of lead in America's gasoline dropped by 46 percent.[42] The great phaseout had begun.

But the Ethyl Corporation wasn't going to go down without a fight. They sued the EPA, claiming that they had been deprived of property rights by the agency's "arbitrary and capricious" lead regulations. Ethyl argued that it wasn't enough for the EPA to base their decisions on "significant risk," but rather that they had to show "actual harm."[43] It is easy to see why a polluting industry might prefer a standard of "actual harm"—its strict definitiveness allows them to undermine pretty much every possible piece of scientific evidence. Since it is impossible—or at least horrifically unethical—to carry out a study in which large groups of people are intentionally exposed to different levels of a toxin, we are left with studies that have inherent imperfections. Animal studies may not perfectly reflect human physiology. Epidemiological studies always have potential confounding factors—income, location, genetics, and so on—that may impact the outcome. The industry wanted to set a standard of evidence so high that it could never be reached, and they could never be regulated.

The U.S. Court of Appeals was having none of it. In 1976, they found that "significant risk" was an acceptable standard, and the EPA's regulation of leaded gasoline was within its authority. The drawdown of lead in gasoline continued, despite the leaded gas industry trying to portray themselves as victims of regulatory overreach. One Ethyl executive was particularly hyperbolic: "The whole proceeding against an industry that has made invaluable contributions to the American economy for more than fifty years is the worst example of fanaticism since the New England witch hunts in the Seventeenth Century."[44] The industry predicted that the phaseout of leaded gasoline would decimate the country's gas supply and irreparably harm the entire economy.

The industry's hopes were raised by the election of President Ronald Reagan, who was elected in 1980 on a business-friendly platform of deregulation. Reagan's first EPA administrator, Anne Gorsuch (whose son is now a Supreme Court justice), let it slip to a visiting gasoline refiner that she would not enforce the current limits on lead in gasoline, and was planning to repeal them soon.[45] Her statements got out, and the backlash was immediate. The comic strip *Doonesbury* dubbed Gorsuch "The Ice Queen" and depicted an EPA staffer refusing to come in off a building ledge "until Gorsuch publicly agrees that the purpose of the Environmental Protection Agency is to protect the environment."[46] By this time, environmentalists and child health advocates were organized enough to be a political force to be reckoned with. Under intense political pressure, the EPA changed course and continued the phaseout. Later, Gorsuch would be forced to resign due to a scandal about the mishandling of Superfund money.[47]

The Reagan administration also actively worked to suppress information about the dangers of leaded gasoline. A congressional act in 1986 mandated the writing of an extensive report on childhood lead poisoning, to be presented to Congress. For six months, researchers Annemarie Crocetti and Paul Mushak worked to draft a 300-page report documenting potential harmful effects of lead at levels well below the 25μg/dL cutoff that the government was then using.[48] However, when the draft was complete, its authors were informed that Congress would only receive a substantially watered-down "summary" that omitted their most serious findings. Both researchers resigned in protest.[49] Only after many months of increasing pressure from Congress did Reagan's Department of Health and Human Services finally release the full report.[50]

Even as Reagan and his appointees were striving to implement their anti-regulatory philosophy, government employees were demonstrating the value of environmental regulations. Joel Schwartz, an economist working at the EPA, was assigned to calculate the economic impacts of the lead phaseout on oil refineries. Schwartz decided that if he was going to calculate the costs of the phaseout, he would calculate the benefits as well. Schwartz and his team estimated the increased health care costs associated with increased lead exposure, and they used Needleman's research into the effects of lead exposure on IQ to estimate the cost of lost education and wages. This project is thought to be the first major cost-benefit analysis of a health policy.[51] The analysis showed that while removing lead from gasoline would indeed cost the oil industry around half a billion dollars, the benefits of such a policy added up to nearly twice that amount.[52]

We now know that Schwartz's analysis greatly underestimated the benefits, because it only considered health and educational effects for children whose blood lead levels were above 30μg/dL.[53] Even some of the Reagan administration's most anti-regulation crusaders were swayed by Schwartz's results. In the end, instead of relaxing standards on lead in gasoline, the EPA tightened them further. The phaseout continued until its completion on January 1, 1996.[54]

By 1986, the amount of lead in American gas had fallen by 90 percent.[55] Leaded gas continued to account for a tiny percentage of the gas sold in the United States—only 0.6 percent by 1995.[56] In 1996, the EPA banned all use of leaded gas for on-road vehicles, and despite the dire predictions of the oil companies, those companies did not go out of business. Although an oil embargo did force gas prices higher throughout the 1970s, prices then fell dramatically during the 1980s, even as the phaseout of leaded gasoline continued. Oil companies did what companies do—they adapted. Refiners now use other technologies, such as thermal cracking, to raise the octane of their gasoline without having to put in any lead. Americans can still buy "premium" high-octane gas at any gas station in the country. (You probably shouldn't, though. Using a higher octane than your owner's manual recommends doesn't actually gain you anything. Only certain high-performance cars truly require the "premium" gas.[57])

Americans and our cars made the switch to unleaded gas with little difficulty. It didn't even take us very long. The United States had gone from its peak usage of lead in gasoline to zero in a matter of only two decades—a single generation. The amount of lead in our gas had risen steadily for half a century, from the 1920s to the 1970s, and then plummeted. The impact of this change on the amount of lead in the bodies of American children was rapid and dramatic.

4

Lead in America's
Children

▬ ▬ ▬ ▬ ▬ ▬ ▬ ▬

In the 1950s and 1960s, as the use of leaded gasoline was skyrocketing, children in the United States were being exposed to more and more lead. However, there's no way to know precisely how much lead American kids had in their bodies before the 1970s. Until the 1960s, testing kids for lead was difficult, painful, and expensive, so it was typically only done in acute cases of lead poisoning.[1] Most of the data from the 1960s and early 1970s came from screening programs in major cities, which were specifically designed to target the kids expected to have the highest lead levels. So we really don't know what the national average was. We do have some clues, though, about just how bad things were. As we discuss historical blood lead levels, don't forget—these days, doctors will intervene if a child's blood lead level is above 5µg/dL and send the child to the hospital if the level is above 20µg/dL. But *acute* lead poisoning, the kind where kids are catatonic or convulsing, unable to eat or in severe gastrointestinal distress, usually doesn't happen until blood lead levels get above 80µg/dL or so. For much of the history of lead poisoning, acute cases are the only ones we have any record of.

Some of the earliest work on childhood lead poisoning was done in Baltimore. In 1933, the charismatic Dr. Huntington Williams became health commissioner in the city. He was known for his five-minute health-related radio broadcasts, with catchy slogans like this one on spoiled food: "When in doubt, throw it out!" Dr. Williams collaborated closely with researchers at Johns Hopkins University and the University of Maryland to study and address pediatric lead poisoning. It is because of the work of Dr. Williams and his collaborators that we have better historical records of lead poisoning in Baltimore than in any other city in the United States.[2]

Between 1931 and 1951, Baltimore recorded 293 cases of acute lead poisoning in children, occurring most commonly around age two, and more often in the summer than the winter.[3] (Children are outdoors more in the summertime, so they're exposed to more leaded gas fumes and contaminated soil. Also, people tend to open and close their windows in the summer, and the scraping of the windows against their lead-painted window frames creates lead dust.) During these years, doctors at Johns Hopkins and the University of Maryland were trained in how to diagnose and treat cases of lead poisoning, and screening programs were initiated to find potentially lead-poisoned children. These programs had the unfortunate effect of earning Baltimore an unwarranted reputation as a hotbed of lead poisoning at a time when these concerted efforts were actually improving the situation in Baltimore. In many other major cities, untrained doctors were still misdiagnosing cases of lead poisoning as cases of tuberculosis and other infectious diseases.

Eventually, other major cities began to look in earnest for cases of childhood lead poisoning as well. A 1965 screening program in Chicago found that between 5 percent and 15 percent of children screened had blood lead levels above 50μg/dL. Screenings in several major cities in the late 1960s found that between 25 percent and 45 percent of young children screened had blood lead levels above 40μg/dL.[4] Kids outside of major urban centers were affected as well. A 1972 study of young children in smaller cities found that 19 percent of them had blood lead levels above 40μg/dL.[5] Because 40μg/dL was still considered the threshold for "undue lead exposure" at that time, most of these studies didn't report any values lower than that.

In an innovative 2010 study, a group of researchers looked at past lead exposure by measuring lead concentrations in the enamel of molars removed during standard dental procedures.[6] When new molars are formed during childhood, lead gets built into the new teeth along with calcium, and past research had demonstrated that the concentration of lead in the core enamel of a tooth is an accurate indicator of the blood lead level of the child whose mouth that tooth was forming in. This study examined teeth taken from 124 adults, most of them Black, who had grown up in Cleveland between 1936 and 1993. The researchers estimated that during the era of peak gasoline lead use, from 1960 to 1975, the average blood lead level in these children was 48μg/dL.[7]

The results of the Cleveland study also suggest that at least two-thirds of the lead in these children came from gasoline, and that kids who grew up in high-traffic neighborhoods had higher blood lead levels, on average, than kids who grew up in low-traffic neighborhoods. This study also demonstrates that the average blood lead levels of kids in Cleveland was rising throughout the

1940s and 1950s, peaked in the mid-1960s (at an average blood lead level almost three times as high as they found in the mid-1930s), and then fell dramatically throughout the 1970s and 1980s. This pattern supports the idea that leaded gasoline was the main source of childhood lead exposure in mid-twentieth-century America.[8]

This collection of local studies from across the country demonstrates that blood lead levels above 40µg/dL were quite common in urban children in the 1960s and early 1970s. In fact, Dr. Clair Patterson, the geochemist and anti-lead advocate we learned about in chapter 3, estimated that in 1965, the *average* American had a blood lead level of 20µg/dL.[9] Since we know that little kids are exposed to more lead than grown-ups (putting those grubby fingers in their mouths) and that they absorb lead more easily into their bodies, it is likely that the average blood lead level of preschool children was even higher than 20µg/dL in the 1960s and early 1970s. However, it wasn't until the late 1970s that lead exposure was measured in a systematic way across the country.

The first nationwide study of blood lead levels began in 1976. Between 1976 and 1980, the federal government carried out the second National Health and Nutrition Examination Survey.[10] This was a nationally representative survey of more than 20,000 people that looked at a number of different indicators of health and nutritional status, including—for the first time—blood lead levels. (The study is typically known as NHANES II. Since the first NHANES study didn't measure blood lead levels, we can think of this one as a sequel—NHANES II: This Time It's Got Lead.) So the late 1970s are the earliest time period for which we have definitive measurements of the blood lead levels of children across America. It's not a pretty picture.

During the late 1970s, the average American preschooler (kids aged 1–5) had a blood lead level of 15µg/dL. Nearly all kids (99.8 percent) had blood lead levels of 5µg/dL or higher. Of the thousands of kids tested, not one single blood sample came back lower than 4µg/dL. Remember that recent research has shown harmful effects at blood lead levels as low as 2µg/dL. American children were all lead-poisoned, and many of them were quite severely lead-poisoned. One in four little kids across the country had a blood lead level higher than 20µg/dL, and more than 4 percent of kids were above 30µg/dL, a level that was known to be harmful even back then.

The NHANES II data also showed that average blood lead levels were already falling during the period of the study (1976–1980) and that this decline was closely correlated with the drop in gasoline lead during the EPA-mandated phaseout of leaded gas.[11] So we can assume that since even more lead had been used in gasoline in the earlier part of the decade, little kids in the early 1970s had higher blood lead levels than this study found in the late 1970s. This suggests that a nationwide average of 20µg/dL or more in the late 1960s and early 1970s is a reasonable estimate.

The amount of lead in our gasoline dropped precipitously between the early 1970s and the early 1990s, and so did the amount of lead in the bodies of our children. The next National Health and Nutrition Examination Survey was carried out between 1988 and 1991 (NHANES III: These Kids Are Millennials). The change that had taken place in barely more than a decade is quite striking. The average blood lead level for preschool children across the country had dropped from 15μg/dL to less than 4μg/dL. While pretty much all of the kids in the late 1970s had blood lead levels of 5μg/dL or higher, only about a third of the kids in this study did.[12] In the 1970s, around 25 percent of the kids had blood lead levels of 20μg/dL or higher, which today would get you sent to the hospital for further testing. By the 1988–1991 survey, only 1 percent of the kids had levels that high. That's a 95 percent drop in the number of kids with this dangerously high amount of lead in their bodies, in only a little more than a single decade. There is good evidence that this precipitous drop in blood lead levels in children was primarily due to the elimination of leaded gasoline. Leaded gas had been the primary source of airborne lead in the United States,[13] and the amount of lead in the air declined by a whopping 95 percent between 1977 and 1994.[14]

The average amount of lead in American children has continued to decline. The most recent NHANES study, carried out in 2007–2010, showed an average blood lead level for preschool children of around 1μg/dL. The number of kids with blood lead levels of 5μg/dL or higher was down to less than 3 percent. Fewer than 0.1 percent of the kids in this survey were above 10μg/dL, and researchers didn't report any levels higher than that, but we can assume that the number above 20μg/dL was extremely small.[15] No new national survey has been carried out since 2010, but smaller studies suggest that kids' average blood lead levels have continued to fall.

To recap, the prevalence of blood lead levels above 5μg/dL (requires intervention) has gone from pretty much every kid in America to only one kid out of every forty, and the prevalence of blood lead levels above 20μg/dL (urgent problem) has gone from one kid out of every four to less than one kid out of every 1,000, as shown in Figure 4.1. The steepest part of that drop happened before 1990, but the decline is still going on today. Every year, the amount of damage that this neurotoxin is doing to the brains of American children is less than it was the year before. The amount of damage that was done to members of Generation X, those of us born in the 1960s and 1970s, is hard to even comprehend from today's perspective. (We'll learn more about exactly how lead harms children's brains in chapter 5.)

The damage from all those years of leaded gasoline wasn't evenly distributed among all American children. Black kids had more lead in their bodies than White kids in every national survey. (As I mentioned in chapter 1, most of these surveys didn't include any racial or ethnic categories other than Black and

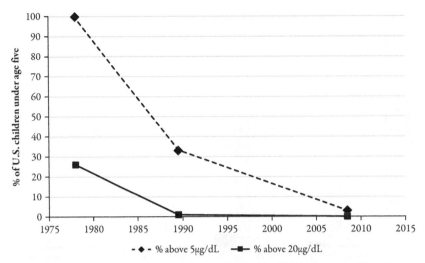

FIG. 4.1 Percentage of U.S. children aged 1–5 with blood lead levels above 5μg/dL and above 20μg/dL over time, based on National Health and Nutrition Examination Survey data.

White. Unfortunately, we don't have good historical data on any other racial or ethnic groups, even though these other groups made up more than 9 percent of the U.S. population by the end of the 1970s.) During the NHANES II study in the late 1970s, when the average blood lead level for White preschoolers was 15μg/dL, the average for Black preschoolers was 21μg/dL. While 18 percent of White kids at that time had a blood lead level above 20μg/dL, a shocking 52 percent of Black kids did. In addition, 12 percent of Black children were above 30μg/dL, a level that was already known to be harmful in the 1970s, compared with only 2 percent of White children. So Black kids were six times as likely as White kids to suffer from the severe damage that comes from having over 30μg/dL lead in your blood.[16]

By the time of the third NHANES study (1988–1991), when only one in four White kids had a blood lead level above 5μg/dL, more than half of Black kids still did. At that time, less than 1 percent of White children had levels over 20μg/dL, but almost 4 percent of Black children did.[17] In the most recent NHANES study (2007–2010), everybody's blood lead levels were much lower, but the average for Black kids (1.8μg/dL) was still almost 40 percent higher than the average for White kids (1.3μg/dL). Also, the chances of having a blood lead level above 5μg/dL were more than twice as high for Black children versus White children (5.6 percent versus 2.4 percent), as shown in Figure 4.2.[18]

There are a number of factors that have contributed to this racial disparity in childhood lead exposure, and certainly differences in average income is one of them. In the NHANES II study in the 1970s, kids in the lowest income

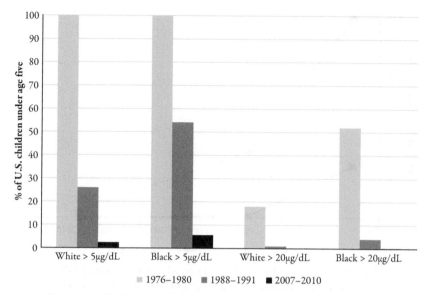

FIG. 4.2 Percentage of White and Black children aged 1–5 with blood lead levels above 5μg/dL and above 20μg/dL in each of the three National Health and Nutrition Examination Surveys.

category had, on average, almost 50 percent more lead in their blood than kids in the highest income category. However, differences in average income by race do not explain all of the difference that studies have found in childhood lead exposure. Even if we look at Black kids and White kids in the same income categories, we still find that the Black kids had more lead in their bodies on average in each of the time periods studied. This racial disparity also remains if we only look at kids whose parents had the same level of education. Something other than just socioeconomic status must account for the different levels of lead exposure by race.

The biggest reason why Black children have always been exposed to more lead than White children in the United States is housing discrimination. Some neighborhoods have more lead in them than others, and a variety of forces have acted together over time to push Black families into higher-lead neighborhoods. There have always been discriminatory housing practices in the United States. By the early twentieth century, a patchwork of laws and regulations limited where Black families could live. Then, in 1934, the Federal Housing Administration (FHA) was created, and it used a set of maps produced by the Home Owners' Loan Corporation to discriminate against majority-Black neighborhoods for decades to come.[19]

"Residential security maps" were created for more than 200 cities across the country, designating different neighborhoods as more desirable or less desirable for the purpose of mortgage lending. The most desirable neighborhoods, outlined in green, were typically the most recently developed areas, which were generally out in the suburbs. The rating system continued through "still desirable" neighborhoods in blue, and "declining" neighborhoods in yellow, and finally to the least desirable neighborhoods, outlined in red on these maps. These were typically majority-Black neighborhoods in the center of cities. Those were the areas that got "redlined."

If private lenders wanted their mortgages to be insured by the Federal Housing Administration, they had to follow the FHA's recommendations, which included a suggestion to avoid these areas with what the FHA called "inharmonious racial groups." The banks' unwillingness to provide mortgages for properties in these redlined areas meant that many families were unable to purchase homes, and were stuck renting from often unscrupulous landlords. Much of the housing stock fell into disrepair, and a lack of resources left these redlined neighborhoods with crumbling infrastructure.[20]

The Federal Housing Administration's recommendations for mortgage lenders also included a preference for racially restrictive zoning ordinances. White neighborhoods explicitly excluded Black families, with zoning laws and neighborhood regulations that specifically prohibited selling to anybody who was not White. Many homes even had this exclusion written into their deeds, which proclaimed that the property could only be sold to White buyers, forever. The result was that Black inner-city neighborhoods became even more homogeneous as White families left for these segregated enclaves in the suburbs.[21]

In many cities, the widespread development of public housing projects, starting in the 1930s, was a boon to families because it provided low-cost housing options. However, public housing programs often had the effect of increasing segregation, grouping even more low-income Black families together in center-city neighborhoods. After World War II, the GI Bill provided subsidized mortgages that allowed returning soldiers and their families to buy houses in unprecedented numbers, leading to new suburbs springing up all over the country. These suburbs continued the practice of excluding Black buyers, often through legal restrictions. In the famous Levittown housing developments, both buyers and renters were specifically required to be of "the Caucasian race."[22] Throughout the 1950s and into the 1960s, government housing policy had the ongoing effect of worsening segregation.

Then came the Fair Housing Act of 1968, which outlawed housing discrimination based on race. Under this law, anyone who explicitly refused to rent or sell to a Black tenant or buyer could theoretically be prosecuted. For prosecution to occur, though, Black prospective tenants and buyers would have to prove in a court of law that they had been discriminated against. Think about what that requires—the money to hire a lawyer, the time to build a case, a flexible work schedule that allows you to appear in court, a judge who takes you seriously, and a jury unbiased enough to believe you. Needless to say, many racist landlords and property owners went unpunished, and, in many cases, overt housing discrimination continued.

During the 1970s, the New York Human Rights Division worked to find cases of discrimination by using "testers," sending Black and White employees to the same building to inquire about apartments available. In one building, a Black tester was told that the apartment had already been rented, and then a White tester went back to ask about the same apartment and, according to her, "They greeted me with open arms." Eventually, a number of companies— including, in 1975, one owned by real estate magnate Donald Trump— reached settlements with the Justice Department, requiring, among other things, that they put ads in newspapers declaring that Black applicants were welcome.[23] Clearly, discriminatory housing practices continued to be common long after 1968.

In fact, discriminatory housing practices are still going on today. In study after study, landlords have continued to show a preference for White renters. For example, in 2011, researchers sent responses to 14,000 Craigslist rental postings in thirty-four cities, using either "White-sounding" or "Black-sounding" names. In the absence of any other information, landlords were significantly less likely to respond favorably to potential renters with "Black-sounding" names.[24] Discrimination affects home buyers as well as renters—as recently as 2017, researchers found that Black home buyers pay more than White home buyers for equivalent properties, and the difference cannot be explained by differences in income or access to credit.[25]

Many people have written about the lasting impacts of housing discrimination on Black communities today. Two-thirds of the accumulated wealth of a typical household is home equity, and being effectively excluded from this avenue for building wealth has made it more difficult for Black families to build up wealth over time. In 2013, a typical White household had managed to accumulate $134,000 in assets, while a typical Black household had only $11,000.[26] In addition, public schools in America are still quite segregated, and majority-Black schools are significantly less well funded than their majority-White counterparts.[27] Black neighborhoods often have fewer grocery stores than White neighborhoods, making it harder to buy healthy foods like fresh fruit and vegetables.[28] And harsh police practices in majority-Black neighborhoods have recently been front-page news.

Childhood lead exposure is just one item on a long list of ways that housing discrimination and segregation have harmed Black Americans. The inner-city neighborhoods that Black families were often forced into have some of the oldest housing units in the country, so they are the most likely to contain lead paint. The prevalence of poverty in these neighborhoods has resulted in housing that is often poorly maintained, with chipping or flaking paint that is especially dangerous for children. The situation has been exacerbated by the discriminatory lending practices described above—it is difficult for landlords in many predominantly Black areas to borrow the money they need to make repairs and upgrades. These older, inner-city neighborhoods also frequently have lead pipes. The use of lead pipes was all the rage in the late nineteenth and early twentieth centuries, when water systems in many of our center cities were installed. Typically, outlying suburbs were built after the use of lead pipes had become much less common.

Black neighborhoods also, on average, were exposed to more of the lead from leaded gasoline than White neighborhoods. Inner cities have always had more traffic than suburban or rural districts—more people in a smaller area means more cars and trucks going by your front door. That discrepancy got even worse with the construction of the interstate highway system. The original proposal for an interstate highway system in the United States focused on "superhighways . . . mainly in metropolitan areas" in order to reduce traffic congestion.[29] In addition, the highway system was intended to promote the national defense, and the military wanted to be able to move troops and weapons quickly in and out of the centers of major cities if they should ever have to fight a war on American soil. Within these cities, Black communities typically lacked the political clout to fend off the siting of highway projects in their own backyards. So, starting in the 1950s, the U.S. built multilane highways through the middle of many Black inner-city neighborhoods, often tearing down houses, businesses, and schools in the process.[30]

In the decades that followed, kids living near these major highways were exposed to the lead coming out of the tailpipes of the thousands of cars and trucks driving through their neighborhoods. Although Dr. Clair Patterson's work showed that small amounts of lead can travel great distances, other scientists have demonstrated that much of the lead doesn't travel very far at all. One study found that almost a quarter of the lead released from a stretch of roadway can be found in the plants and soil within one hundred meters of the road.[31] A hundred meters is about the length of a football field. During the era of leaded gasoline, kids living right near high-traffic roads were getting a much higher dose of lead than kids living even half a mile away. Years of discrimination, segregation, and racist urban planning meant that an awful lot of those kids living near high-traffic roads were Black kids.

It is easy to forget how much our lives today are shaped by historical forces. Where you live now is influenced by where your grandparents lived. This influence can be direct—many people like to live near family—or indirect. If your grandparents were able to build up home equity over time, perhaps they could help your parents, or even you, with financial support that expanded your geographic opportunities. Where your grandparents lived was affected by legal, government-sanctioned racial discrimination until 1968, and by illegal, but still widespread, racial discrimination right up to the present day.

The amount of lead that was in your environment during your baby and toddler years was largely determined by where you lived. A neighborhood with brand-new homes, or expensively refurbished older homes, doesn't have lead paint chipping off the baseboards and window frames. A neighborhood in a recently developed suburb doesn't have lead pipes. Also, until the 1990s, the amount of lead in your childhood environment was largely determined by how much traffic was moving through your neighborhood on crowded city streets and urban superhighways.

The amount of lead in your environment during your critical early years determined how much lead got into your body. Everybody breathes the air, and all babies and toddlers put their hands in their mouths. Some doctors once argued that certain inner-city kids only ingested lead because they weren't properly supervised, but this argument is clearly not valid. Hand-to-mouth behavior is common even in babies who haven't been born yet, as seen in pregnancy ultrasound images. Studies have shown that this behavior continues to be common among children for the first several years of life.[32] So it doesn't really matter who was keeping an eye on you—a parent, another family member, a paid caregiver; if there was lead in the dust and soil around you, you were eating it. Housing discrimination has led directly to racial disparities in the amount of lead in children's environments, which has inevitably caused racial differences in the average amount of lead in children's bodies.

These differences have real and lasting consequences. The development of your brain—the very brain that is making it possible for you to read these words right now—was influenced by how much lead was in your body, from the time you were conceived through the first several years of your life. Unsurprisingly, changes in brain development can have a profound influence on people's chances of academic success and their behavior throughout their lifetime.

5

Brains and Behavior
and Lead

▬ ▬ ▬ ▬ ▬ ▬ ▬ ▬ ▬

In order to understand the impact that lead has on developing brains, it's helpful to know some details about how brains develop. Let's take a look at some
of the highlights of brain development from conception through adolescence.
When you were first conceived, you started out as a hollow ball about the size
of the period at the end of this sentence. Then that ball folded in on itself until
it turned into a hollow tube, with three layers of cells. (Early embryonic development involves a lot of folding, like origami.) The outer layer included the cells
that would become your nervous system. Those cells received numerous chemical signals—including one chemical that scientists refer to as "noggin"—and
underwent some more folding, and by about a month after egg met sperm, you
had the basic structure that would eventually become your brain.[1]

During the rest of your time in utero, you were constructing your neurons,
the cells that carry messages in your brain and the rest of your nervous system.
You made new neurons at an incredible rate—more than 5,000 of them every
second. By the time you were born, you had as many neurons as you would need
for your entire life—around 100 billion of them—and those neurons were hard
at work getting organized. They arranged themselves in all the different parts
of your growing brain, and got into position to communicate with each other
using cellular structures called axons and dendrites.[2] (You can see these structures in Figure 5.1.)

A neuron is a unique type of cell; to understand its structure, imagine a scale
model of one neuron as a smallish octopus. Seven of its tentacles are each about

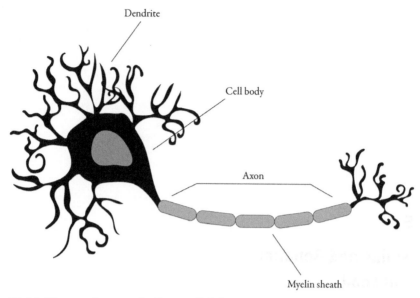

FIG. 5.1 Diagram of a neuron, by Shawntrell Coleman.

a foot long—those are the dendrites, all clustered around the main part of the cell. (A real neuron can have far more than seven dendrites.) These dendrites are responsible for receiving signals from other nearby cells. The eighth tentacle of our scale model octopus is the axon, and relative to those foot-long dendrites, it could be ten feet long, or as much as two miles long! The axon carries the message received by a dendrite to another cell, which could be adjacent or could be some distance away.[3]

So the dendrites receive signals from other neurons, and the axon passes those signals along to the next neuron down the line. The signals moving down the axon are electrical impulses, and they move more efficiently if the axon is coated with an electrical insulator, just as the power cord of an electronic device is coated with rubbery plastic. The substance that coats your axons is called myelin, so the process by which your neurons acquire this insulating layer is called myelination. Myelination is critical for successful brain development because nerve impulses can't be passed along efficiently enough to generate complex thoughts and behaviors until that insulation is in place. Your brain began its process of myelination around the start of the third trimester, and the process picked up speed when you were a baby, helping you transform from a relatively inert newborn into a walking, talking toddler. Myelination continued all the way up to adolescence, and was basically done by the time you were in middle school.[4]

Back when you were born, you had an enormous number of neurons, each one able to connect to lots of other neurons. The connection between two neurons is called a synapse; it's a tiny space that the signal travels across to get from one neuron to another. After you were born, you went through a period of "synaptic exuberance," forming tons of new connections. The dendrites of each neuron were reaching out, finding and hooking up with any nearby axons. By the time you were two, you had about twice as many synaptic connections as you would eventually need. The next phase of brain development was focused on pruning away the excess synapses. A lot of learning early in life involves the process of strengthening the useful connections and getting rid of all the unnecessary, distracting ones. Throughout your childhood and into adolescence, your brain was eliminating excess synapses, turning itself into a lean, mean thinking machine.[5]

Your nine months in utero and your first few years of life were the most critical period for all of this brain development. A lot of things had to go right for you to end up with a healthy, functional brain. Nearly everything that neurons need to do during development—grow, form synapses, fine-tune synaptic connections—is dependent on the element calcium. When the dendrite of one neuron first encounters the axon of another neuron, calcium is released inside the dendrite to alert the cell that contact has been made. Once the synapse is established, calcium helps to regulate the transmission of signals from one neuron to another. Over time, calcium plays an important role in helping to determine which connections stick around and which ones are pruned away. Anything that interferes with the ability of calcium to do its job in the brain has the potential to do a lot of damage during those early years.[6]

As an element, lead has a very unfortunate tendency to act similar to, but not quite exactly like, calcium in the body. Calcium can send signals to brain cells by binding to particular proteins, like a key in a lock. Add calcium, and these proteins turn on and carry out their function; when calcium levels drop, the proteins turn off again. For some of these proteins, lead atoms bind even more effectively than calcium atoms, like a key that gets jammed in the lock, so the function turns on, but can't turn off again. Cells need to be able to turn these protein functions on and off, just as a thermostat turns the furnace on and off to keep the house at the right temperature. When lead gets jammed in where calcium should be, the proteins can act like a broken thermostat, running the furnace day and night. The house would be far too hot, and eventually the furnace would break down. In the same way, neurons don't work properly if they have lead atoms stuck where calcium atoms should be. They aren't as good at making connections with other neurons, or at prioritizing important connections over extraneous ones.[7]

Lead doesn't just alter the function of neurons; it can also mean life or death for them. Normal brain development involves making more neurons than you need, and then getting rid of the ones that aren't making useful connections. So all the neurons in your brain are born with a mechanism that allows them to, basically, commit suicide. This mechanism is, like so many others, based on calcium signaling. Lead can get in there and mess up the signaling, leading to the death of important neurons and the survival of extraneous neurons that are just getting in the way. Plus, lead can kill the cells that build the insulating myelin layer around the axons. During the first few years of life, the brain is busy making neurons, connecting those neurons, pruning away extraneous connections, and assembling a protective coating of myelin. Every one of those critical processes can be disrupted by the presence of lead.[8] All of these harmful impacts of lead on the developing brain are made possible because lead is able to pass through the blood-brain barrier. This highly selective barrier keeps some other toxic chemicals from reaching the brain, but lead's chemical similarity to calcium allows it to sneak by.[9] Although lead can be cleared from the brain in a matter of months if no new lead is entering the body, the damage it has done to the brain's development is permanent.

The brain isn't the only place in the body where lead can cause harm. Lead has also been shown to harm the immune system, the cardiovascular system, kidney function, and reproduction. As with the harm to the brain, it doesn't take a whole lot of lead to cause these problems throughout the body—all of these negative effects have been shown at blood lead levels of 5µg/dL or lower. In the immune system, higher levels of lead can kill the cells that patrol the body and fight off invading bacteria and viruses. At lower concentrations, lead doesn't actually kill many of these cells, but it alters them in ways that make them worse at doing their jobs. This is especially insidious, because if you're having trouble fighting off infections, a doctor might test how many immune cells you have circulating in your blood, and that number could come back in the normal range. The doctor wouldn't realize that even though the cells are there, they aren't working properly because lead has damaged their ability to function.[10]

Having lead in the body also raises the blood pressure, and the higher the concentration of lead, the greater the elevation in blood pressure. This is bad news because high blood pressure puts you at greater risk for heart attack, stroke, and other cardiovascular problems. Numerous studies have shown that lead exposure increases your chances of having a cardiovascular disease and/or increases your chances of dying from that disease. One 2018 study estimated that over 250,000 deaths each year in the United States from cardiovascular disease are attributable to lead exposure.[11] There are a number of possible explanations of why lead might raise blood pressure—it changes the ways that cells send signals to each other, and may also inhibit your body's natural antioxidant powers. Many scientists think that the primary reason that lead makes your blood pressure go up is because of the harm that it does to your kidneys.[12]

Kidneys are responsible for managing the amount of water in your body—when you're well hydrated, they remove some of the water circulating in your bloodstream and turn it into urine to be eliminated. Lead kills the cells in your kidneys that are responsible for this function, so lead-exposed kidneys can't filter water out of the blood as fast as healthy kidneys do. This can cause high blood pressure, but that's not all. Kidneys are also responsible for filtering toxins out of your blood, and lead makes them worse at this as well. This can allow harmful chemicals to build up to dangerous levels in your blood. Chronic kidney disease can result in the need for dialysis or a transplant, and can even be fatal if not treated.[13]

Exposure to lead also harms the reproductive systems of both men and women. Men with higher levels of lead exposure produce fewer sperm, and the ones they do produce don't swim as well, on average, as healthy sperm. In women, lead harms cells in the ovaries and fallopian tubes, reducing the chances of conception. If pregnancy does occur, higher lead exposure increases the chances of miscarriage or low birth weight. During pregnancy, some calcium is released from the mother's bones in order to provide enough calcium for the growing baby, and lead can be released along with that calcium. So when a woman gets pregnant, some of the lead that was stored in her bones many years ago can get mobilized and start circulating in her bloodstream, and in the baby's bloodstream. This process of lead being released from the bones continues during breastfeeding as well, exposing the babies of lead-poisoned mothers to some of that old lead.[14]

Nearly all the cells and organs in your body are potentially susceptible to harm from lead. Some of the problems caused by lead are a result of its tendency to mimic calcium, which is an important element everywhere in your body, and some problems are caused by other nasty things that lead atoms can do. No part of you is safe. So, as we think about the rise and fall of leaded gasoline, it's worth remembering all the different adverse health effects that lead can cause. As nationwide lead exposure plummeted in the 1980s, we got healthier in many ways. The drop in lead levels benefited adults as well as children. The harm that lead does to our immune and cardiovascular and reproductive systems, and to our kidneys, affects people of all ages. So less lead in the environment is good news for your whole body, no matter how old you are.

It's clear, though, that brains were the biggest beneficiaries of the drop in lead levels, and that children's brains benefited most of all. Harm to the developing brains of children is the health effect of lead that has been the most consistently demonstrated at the lowest levels. Also, brain cells aren't good at regenerating themselves, unlike many other types of cells in the body, so damaging brain cells during early childhood can result in lifelong injury to the brain. That's why so much lead research is focused on children. While even low levels of lead can potentially do some damage in adults, the damage to children's brain development is the most severe, well-documented, and lasting harm that is caused by exposure to lead.

This damage done by lead in the brain can be seen at the cellular level in animal studies. In one study in Germany, researchers put electrodes into the brains of rats to measure something called "long-term potentiation" (LTP).[15] LTP is a measure of the strengthening of the synapses that are being used the most. As you can imagine, this process is very important for learning and over-all brain function. For example, earlier in this chapter, you read about what synapses do. When you were reading that section, the pattern of connections in your brain that relate to the concepts of "synapse" and "connection" and "neuron" got stronger. Now that you're reading about synapses again, that strengthening of connections in your brain is helping you remember what we're talking about. The ability of your synapses to undergo long-term potentiation is helping you understand this chapter, and will help you to remember it in the future. So that's quite an important thing for your brain to do.

In the German study with the rats, some of them had been given a constant low dose of lead in their drinking water since birth, and some had been given clean water. The rats who had been exposed to lead had weaker LTP than the rats who hadn't been exposed. The lead harmed the ability of their brains to strengthen the connections that were being used the most. Since we know from many other studies that LTP plays an important role in learning and memory, this means that lead impacts the cells of the brain in ways that can have seri-ous consequences. Importantly, we know that it was lead exposure that caused the difference in LTP between the two groups of rats, and not some other factor, because the researchers ensured that everything else in their environment, except for the amount of lead in their drinking water, was the same for both groups.[16]

Since human research subjects might not be enthusiastic about having elec-trodes inserted into their brains, it's more difficult to measure the effects of lead on the human brain at the cellular level. However, scientists have looked at the effects of lead on the overall anatomy of human brains, using advanced brain scanning techniques. A number of studies have shown negative impacts of lead on human brain structure, including the Cincinnati Lead Study, which fol-lowed children from inner-city neighborhoods, starting before they were even born.[17] In the late 1970s and early 1980s, the study recruited pregnant women in Cincinnati and monitored the lead levels of their kids while they were still in utero, and then throughout their childhoods. In an impressive feat of scien-tific persistence, these researchers kept track of their research subjects for over two decades and managed to convince 157 of them to spend time in a claustro-phobic, noisy machine so their brains could be scanned and measured.

The study used MRI (magnetic resonance imaging) to measure the amount of gray matter in different parts of the brain. Gray matter consists mainly of neurons, so it's the most active component of the brain. What the researchers found is that the more lead a person was exposed to in childhood, the less gray matter they had in certain areas of their brain. This loss of brain volume in lead-exposed subjects was especially dramatic in the prefrontal cortex.[18] If you touch your forehead, then under the skin and under the hard skull, the part of your brain closest to your fingertips is your prefrontal cortex. It's basically the control center of your brain—it organizes your thoughts, feelings, and actions. The prefrontal cortex is important for attention, memory, empathy, and even dreaming. Past research has suggested that bigger is indeed better when it comes to this brain region—in general, people with a larger prefrontal cortex perform better on a variety of cognitive tasks than people with a smaller prefrontal cortex.[19] So, if the growth of a person's prefrontal cortex is stunted by childhood lead exposure, that's really bad news.

The tangible impacts of the damage done to brains by lead have been shown in study after study. Before we look at some of those studies in detail, I want to talk about a couple of big ideas that are important to keep in mind in examining any kind of research on human brains and behavior. The first big idea is that these studies measure *aggregate* effects. If you take a whole bunch of kids who were exposed to lots of lead, give them a test, and average their results, and then compare that average with the average test score from a bunch of kids who weren't exposed to so much lead, you'll generally find a statistically significant difference. However, that doesn't mean that you can predict the outcome for any *individual* person taking that same test. There's a lot of variability. So, on a particular test, you'll still have some kids with high lead exposure who do really well, and some kids with hardly any lead exposure who bomb the test.

An analogy that I find helpful in thinking about these kinds of aggregate effects is the relationship between smoking and lung cancer. We know that the more cigarettes you smoke, the higher your risk of developing lung cancer. Even so, there can be nonsmokers who come down with lung cancer and lifetime heavy smokers who never do. Researchers can make pretty accurate predictions about *populations*—for example, if smoking rates are lowered by X amount, they would predict a reduction in lung cancer diagnoses of Y amount. On the other hand, doctors can't necessarily make accurate predictions for *individuals*. It is impossible to know whether any particular smoker or nonsmoker will eventually contract lung cancer.

In the same way, we can't look at the amount of lead that one particular kid was exposed to and predict the outcomes for that particular kid. The reason for this brings us to the second big idea—any outcome as complex as thinking or behavior is necessarily influenced by *multiple factors*. Lead exposure influences how your brain develops, but so do genetics, home life, education, nutrition, and many, many other things. The smoking analogy is helpful here, too. For years, cigarette companies tried to use the existence of multiple factors to undermine the link between smoking and lung cancer. Lung cancer isn't caused by smoking, they argued, it's caused by genetics, or by eating habits, or by air pollution. Clearly, those factors can indeed affect lung cancer rates, but that doesn't mean that smoking *isn't* a factor. A single outcome can have more than one cause.

Similarly, lead exposure is only one of many, many factors that influence thinking, learning, and behavior. It can be difficult to tease apart these factors, because many of them are correlated with each other. Kids who are exposed to a lot of lead are often also kids from low-income neighborhoods who attend underresourced schools and have parents without much higher education. If we gave a test to only two kids, one from a high-lead-exposure neighborhood and one from a low-lead-exposure neighborhood, we wouldn't know which of these various factors was influencing those two kids' test scores. The way to measure the effect of lead in the face of multiple correlated factors is to test many, many kids, and then use statistical techniques to subtract out the effects of as many other factors as possible.

These statistical methods aren't perfect. There could always be some other factor that we're not aware of. Maybe kids with high lead exposure happen to also be kids who like lollipops, and lollipops actually alter thinking and behavior. We can never account for *everything*. But these methods do a pretty good job of separating out the effect of lead from the effects of other known factors. This allows us to ask the question, "For kids of the same age, the same race, and the same socioeconomic status, does higher lead exposure result in different outcomes, on average?" In study after study, the answer to that question is yes.

Back in the 1970s, researchers had already shown that exposure to lead in childhood is associated with reduced IQ. In the early 1970s, the Centers for Disease Control sent pediatrician Dr. Philip Landrigan to study a lead smelter in El Paso, Texas. Dr. Landrigan published a study in 1975 in a highly respected medical journal, *The Lancet*, showing that kids with blood lead levels above 40μg/dL had significantly lower "age-adjusted performance IQs" than kids with lower lead levels. The low-lead group had been intentionally chosen to match the high-lead group in terms of age, sex, socioeconomic status, and language spoken at home.[20] In 1979, Dr. Herbert Needleman published a study in another prestigious journal, the *New England Journal of Medicine*. As we learned in chapter 3, Dr. Needleman's study used baby teeth to estimate lifetime lead exposure, and found that the children with high lead exposure scored lower on IQ tests than the children with low lead exposure. This difference held up even when Dr. Needleman used statistical techniques to subtract out the effects of thirty-nine other factors, including parental education, income level, and so on.[21]

Now, the whole concept of IQ (intelligence quotient) is controversial. Questions used on IQ tests have been shown to be culturally biased, creating a disadvantage for subjects whose cultural background is different from that of the researchers.[22] Many have argued that human beings possess a number of different types of intelligence, which cannot be accurately identified with a single IQ score.[23] And IQ tests have historically been used in some horrific ways, including the forced sterilization of low-scoring individuals as recently as the 1970s.[24] In addition, IQ testing has been used to back up some highly dubious, often racist, arguments.[25] There is no question that the idea of measuring a person's "intelligence quotient" is problematic.

Personally, I would never advocate using an IQ test as the primary way to evaluate a particular individual. However, I would argue that measuring impacts on IQ is a valid way to examine the effects of childhood lead exposure in the *aggregate*. For one thing, these studies typically subtract out the effects of race, socioeconomic status, and language spoken in the home—all factors that help create the cultural bias inherent in IQ scores. Also, these studies are looking at whole populations, not individuals. When we're talking about an individual person, other factors may be more important than a simple IQ score, but when we're examining whole populations, IQ scores can predict differences in a number of outcomes—in education, in life—that we would consider important. Here's another analogy: a kid shouldn't use height as the only factor when picking a classmate to be on their basketball team. That kid may have a short classmate with fantastic ball handling skills, or a tall classmate who is athletically inept. But it is true that, on average, great basketball players are taller than the general population.

As long as we remember that we're talking about aggregate effects, not individuals, it's worth noting that higher IQ scores have been shown to predict success in school and in the workplace. On average, people with higher IQ scores are likely to make more money, commit fewer crimes, and live longer than people with lower IQ scores.[26] So, researchers may not agree about exactly what IQ tests are measuring, but when a study finds that childhood lead exposure reduces IQ, that suggests that lead is causing real, significant harm that is likely to impact people's lives in a negative way. Indeed, since the 1970s, studies have continued to find that more exposure to lead during childhood is consistently associated with lower IQ.

A major review article was published in 2005, written by fourteen experts from a dozen different hospitals and universities. They looked for the highest-quality studies of lead and IQ, the ones that followed kids from birth or infancy until elementary school, measuring lead exposure and IQ at multiple points along the way. The authors of this review looked at the results from seven different studies that met their criteria, and consistently found that kids with higher average blood lead levels had lower average IQ scores. This trend existed all the way down to the lowest blood lead levels they measured—there was no "safe" level of lead. Raising kids' blood lead level from 0μg/dL to 5μg/dL was found to cost them around three or four IQ points, on average. Raising their blood lead level from 0μg/dL to 20μg/dL reduced average IQ scores by about ten points.[27] Numerous other studies have found similar results.

These reductions in IQ score may not seem all that large. After all, a person's IQ measurement might change by three or four points from one day to the next, or from one test to another. However, when these changes are applied to whole populations, they can make a huge difference. IQ scores fall along a bell curve—lots of people with scores near the average, but only a few people with very low or very high scores. If you shift the entire curve downward by four points, it doubles the number of people considered to be "mentally impaired" and cuts in half the number of people in the "genius" range. Even a relatively small shift in average IQ can make a big difference for an entire generation. Remember that IQ has been correlated with income, happiness, and even longevity, so these shifts represent real impacts on quality of life.

The effects of childhood lead exposure show up on real-world standardized test scores as well. Economist Dr. Jessica Wolpaw Reyes found that kids who had higher blood lead levels as preschoolers went on to perform worse on a third and fourth grade standardized test in Massachusetts than kids whose blood lead levels had been lower.[28] Other researchers have found a connection between preschool blood lead levels and performance on fourth grade reading and math evaluations in North Carolina[29] and in Milwaukee, Wisconsin.[30] A study in Avon, England, found that blood lead levels at age two had an impact on standardized test scores at age seven or eight.[31] Similar results have been found in Taiwan[32] and Mexico.[33] Whether we look at IQ tests taken in a laboratory or academic tests taken in the classroom, the research consistently finds that blood lead levels in early childhood are linked to performance later on.

Unfortunately, these effects of childhood lead exposure don't go away as the affected children grow up. One of the best long-term studies of childhood lead exposure was published in 2017 in the *Journal of the American Medical Association*.[34] This study involved a group of individuals born in New Zealand in 1972 and 1973. These kids had their blood lead levels tested at age eleven, and the results ranged from 4μg/dL to 31μg/dL. New Zealand didn't ban leaded gas until 1996, so this group had been exposed to fumes and dust from leaded gas throughout their childhoods.

Amazingly, researchers were able to track down more than 500 of these kids when they reached the age of thirty-eight. The research subjects who had more lead in their blood at age eleven had, on average, lower IQs as adults, even when researchers subtracted out the effects of other factors like their parents' IQ scores. The researchers also looked at socioeconomic status, as represented by occupation. In general, the lower the childhood blood lead level, the higher the subjects scored on the socioeconomic status scale. Because they had been following these folks since childhood, the researchers were able to measure social mobility. On average, the kids with less than 10μg/dL of lead in their blood grew up to have a higher socioeconomic status than their parents, while kids whose blood lead level measured more than 10μg/dL grew up to have a lower socioeconomic status than their parents.[35] Lead exposure doesn't just reduce scores on IQ tests—these results suggest that the damage that lead does in childhood significantly impacts job prospects later in life.

Intelligence—however we define or measure it—isn't the only outcome that's affected by lead. Kids with higher lead exposure are also more likely to exhibit attention deficits and hyperactivity. In addition to looking at IQ, Dr. Herbert Needleman's groundbreaking study in 1979 asked the children's classroom teachers to complete an eleven-question survey about each child. The questions included things like "Is this child easily distracted during his/her work?" and "Do you consider this child hyperactive?" The kids with higher amounts of lead in their baby teeth were more likely to have undesirable answers from their teachers on all eleven questions.[36]

More recent studies have shown the same result—childhood lead exposure is associated with attention deficit/hyperactivity disorder (ADHD). Researchers working in a number of different universities and hospitals, in the United States and abroad, have found a consistent relationship between childhood blood lead level and the chances of being diagnosed with ADHD. Even low levels of lead exposure increase the risk of ADHD—one study only looked at kids with blood lead levels below 3µg/dL, and still found that on average, more lead meant more ADHD.[37] Attention deficit/hyperactivity disorder has been associated with both academic and behavioral problems. In fact, several studies have found that the aspect of ADHD that is most influenced by lead exposure is "impulsivity"—the tendency to act without carefully considering the consequences.[38] Considering all the ways that increased impulsivity can influence a person's education, employment, and relationships, this clearly indicates the potential for enormous lifelong impacts of childhood lead exposure.

Whether or not they have been diagnosed with ADHD, kids with higher lead exposure are more likely to exhibit behavioral problems. This effect starts early in life. One study looked at one- to three-year-olds, and already the kids with higher blood lead levels were more likely to exhibit "hyperactive/distract-ible/easy-frustration behaviors," as identified by trained experts.[39] Another study, using reports from classroom teachers about the behavior of their first grade students, also found that the kids with higher lead exposure had more behavior problems.[40] A nationwide study found a significant link between lead exposure and conduct disorder in older kids.[41] In all of these studies, researchers subtracted out the effects of a number of other factors, such as socioeconomic status and family situation, that might affect children's behavior. The researchers consistently found an effect of lead exposure itself, separate from these other factors.

As we might expect, when kids get older, the ones with higher lead exposure are more likely to get into trouble, sometimes serious trouble. When researchers asked eleven-year-olds to answer thirty questions about their participation in different types of "delinquent" behavior, the kids with higher lead exposure reported more transgressions than the kids who had been exposed to less lead.[42] In 2010, researchers at the University of Southern Mississippi carried out a review of fifteen different studies that looked at lead exposure and behavioral problems in more than 8,000 kids and teenagers. Their analysis found a significant relationship between lead exposure and "conduct problems," including "aggressive and violent behavior" and "delinquent, antisocial, and criminal behavior."[43]

All of these problems are predictable consequences of the harm that lead does to the developing brains of young children. Lead impairs brain cells and changes brain anatomy in ways that damage learning, memory, and attention. We can measure the results of these changes on different tests of cognitive ability, and see the impacts on behavior in the lab, the classroom, and at home. Although the outcomes of lead exposure for any one individual are hard to predict, we can make general predictions across populations. And we would expect that, on average, members of a population with reduced academic success and diminished impulse control would also be more likely to commit violence and other acts of antisocial behavior.[44]

6

Lead and Violence

Before we get to the human propensity for violence, let's talk about hamsters and cats. Animals aren't humans, of course, and just because something makes hamsters or cats more aggressive doesn't necessarily mean it will make humans more aggressive, though we're similar enough to speculate that it might. The valuable thing about animal studies is that the researchers can control all the variables. The animals who were exposed to lead lived under exactly the same set of circumstances as the animals who weren't exposed to lead, except for the lead itself—there are no other factors to try to subtract out.

A 1999 study in Massachusetts tested the effect of lead exposure on the aggressiveness of golden hamsters.[1] Male hamsters were exposed to low levels of lead from the time they were in utero until they reached maturity. Both the lead-exposed hamsters and lead-free control hamsters were tested for aggressiveness in early adulthood. In the test, an unknown "intruder" hamster was placed in their cage, and researchers observed their response. For the most part, the lead-free hamsters took it easy; they "spent most of their time avoiding the intruders." In the lead-exposed group, on the other hand, "every litter contained individuals that attacked readily and bit their intruders." There were statistically significant differences in swiftness of attack, number of attacks, swiftness of biting, and number of bites. The hamsters that grew up with lead in their systems were noticeably more aggressive toward intruders.

In 2003, a research team in New Jersey tested the effect of lead exposure on the part of the brain that controls aggressive behavior in cats.[2] The researchers placed electrodes into the brains of the cats in a spot where an electrical current would stimulate a predatory attack on a mouse. Then they measured how much electrical stimulation it took to get the cats to attack. When the cats had lead added to their diets, they attacked significantly more easily than lead-free cats did, suggesting that exposure to lead had changed their brain function in a way that made them more prone to aggression. So we know that lead exposure can increase aggressiveness in hamsters and cats.

A substantial body of research suggests that lead exposure increases the chances for violence and other types of criminal behavior in humans as well, but before we get to that, we need to take a more careful look at what we mean by "violence," "crime," and "violent crime." The idea of "crime" is especially problematic. The United States has a long history of defining what is and isn't a crime in ways that scapegoat and victimize marginalized groups. One person's vodka martini isn't a crime, but another person's marijuana is (in most states). When the big pharma companies sell you opioids, that isn't a crime, but when a guy on a street corner sells you the same drug, that is a crime. These differences aren't random or accidental. A high-ranking member of the Nixon administration admitted that the War on Drugs was designed specifically to go after "blacks" and "hippies."[3] We cannot talk about "crime" as though that's a straightforward, value-neutral category of behaviors.

Criminal laws are written in discriminatory ways, and then enforced in discriminatory ways as well. I have personally benefited from the biases of our criminal justice system. When I was thirteen—in 1989, just as politicians were scrambling to see who could be the most "tough on crime"—I was arrested for shoplifting at a local discount store. My parents grounded me, but the legal system went quite easy on me. The judge said that as long as I stayed out of trouble until I was eighteen, my record would be expunged, and it would be as though this incident had never happened. As a teenager in 1989, I had not yet heard the term "White privilege," so it wasn't until years later that I realized how lucky I had been. If I had been a Black teenager in the inner city who was caught stealing, I most likely would have entered an increasingly overburdened juvenile justice system, and that one mistake might have had monumental consequences for the trajectory of my life. For me, as a privileged White person, it was a tiny speedbump. I have benefited from many years of being legally allowed to answer "no" to the job application question "Have you ever been convicted of a crime?"

My experience of White privilege in the U.S. criminal justice system is far from unique. Studies have shown that every level of our criminal justice system exhibits bias against poor people and people of color, from the arresting officers to the lawyers, judges, and juries, to the prison officials and parole boards.[4] So, as we examine the effects of childhood lead exposure on violent crime rates, we cannot necessarily use "arrested," "convicted," or "incarcerated" as accurate proxies for "committed a crime," because at every step in the criminal justice system, some people are more likely than others to be let off the hook.

It's also important to remember that, even among criminologists, there is controversy about what does or does not count as "violent crime." Criminologists Franklin Zimring and Gordon Hawkins point out the challenge of agreeing on a strict definition: "Is a fistfight between eleven-year-olds in a schoolyard problematic violence? If not, what about a fistfight between their fathers in a pub?"[5] So we need to remember that "crime," "violence," and "violent crime" are problematic categories that have historically been applied in inconsistent and unjust ways. Nevertheless, violent crime—however it is defined—is one of the most important issues in any society, affecting victims, perpetrators, bystanders, and everybody who lives in the community where a violent act takes place.

Countless studies have investigated factors that affect rates of violent crime over time, and these studies have clearly demonstrated the influence of many, many different factors. One report from the National Research Council, for example, examined the effects of demographics, the economy, policing, abortion, firearms, drugs, gang activity, and social services.[6] Several major studies of influences on violent crime rates have developed conflicting lists of "significant" factors.[7] One recent review article lists twenty-four different potential causes for declining crime rates, from the predictable—"tighter gun control laws"—to the less obvious—"Internet/media home entertainment."[8] The author, criminologist Dr. Maria Tcherni-Buzzeo, highlights nine of these twenty-four possible explanations as "promising," including "reductions in lead exposure." She points out that a number of factors, including lead exposure, all work together to affect a person's level of self-control, and therefore that person's likelihood of committing a crime.

However, many prominent criminologists have been quick to dismiss the impact of changing levels of childhood lead exposure on changing rates of violent crime. In many cases, researchers who focus on one particular factor affecting crime rates will categorically dismiss other factors as unimportant. We'll examine this important debate in the field of criminology more fully in the conclusion. Of course, rising and falling crime rates are important to everybody, not just criminologists, and many Americans have no idea that one of the factors influencing the spike in violent crime in the late twentieth century was the damage done to developing brains by pollution from leaded gasoline.

Many otherwise well-informed people that I've talked to while working on this book had never heard about the connection between lead and crime, and some who had heard of this connection expressed skepticism. I believe that it's important for everybody to understand the impact that this toxic heavy metal, with its concentration in our environment rising and falling over time due to policy decisions made by the federal government, has had on the threat of violent crime that influences all of our lives. Let's take a look at some of the large number of studies, using a variety of different methodologies, that have found significant effects of childhood lead exposure on later incidence of violent crime.

In a study published in 2002, Dr. Herbert Needleman (the pediatric psychiatrist we met in chapter 3) and his collaborators identified 195 teenage boys in the Pittsburgh area who had been convicted of a crime in juvenile court.[9] They recruited a second set of boys who had not been convicted of a crime, but were otherwise similar to the convicted boys in terms of race, socioeconomic status, and so on. The researchers wanted to "minimize undetected delinquents" in this control group, so they eliminated any subjects with a past court record or a high level of "self-reported delinquency." They ended up with two groups of teenagers who had grown up in the same neighborhoods, attended the same schools, and faced the same socioeconomic challenges. The big difference is that one group had been convicted of a crime, and the other group hadn't.

Dr. Needleman and his colleagues used X-rays to measure the amount of lead in the leg bones of each boy. Remember that lead can stay in the bones for many years, so X-rays allowed the scientists to estimate each teenager's overall lifetime exposure to lead. (This X-ray technique wasn't available back in the 1970s when Dr. Needleman did his earlier research on lead in baby teeth.) What the researchers found was remarkable. The teenagers who had been convicted of a crime had, on average, seven times as much lead in their bones as the teenagers who hadn't been convicted of a crime. Even after adjusting for nine other factors—including race, family structure, parental education and occupation, and so on—their results suggested that kids with high lead exposure are three times more likely to be convicted of a crime than kids with low lead exposure.[10]

Additional evidence comes from the Cincinnati Lead Study mentioned in chapter 5.[11] This study followed about 300 babies born between 1979 and 1984. Their blood lead levels were measured at birth and throughout their first six years of life. When the subjects were in their teenage years, researchers caught up with 195 of them and surveyed them about "offenses against property, persons and other illegal activities such as driving without a license or disorderly conduct."[12] Whether the researchers looked at lead exposure in the prenatal period, or at age six, or averaged across the entire childhood, they found the same result—kids exposed to more lead grew up to be teenagers who committed more "delinquent acts." And the effect was not small. The kids who had the highest levels of lead in their bodies at the age of six reported having committed an average of 4.5 more delinquent acts in the previous year than the kids with the lowest six-year-old lead levels.

Actual arrests followed the same pattern as these self-reported offenses. In 2005, when the kids in the Cincinnati Lead Study were in their early to mid-twenties, researchers used court records to calculate arrest rates for 250 of the original research subjects.[13] The researchers looked at arrests that occurred after the subjects turned eighteen and ignored minor motor vehicle and pedestrian offenses. The authors of the study acknowledge that there are problems inherent in using arrests as a measure of criminal behavior, but they make the case that arrests are a more accurate measure than convictions because of the many factors affecting conviction that have little to do with the original crime. Racial bias in arrest rates doesn't play a major role in this study because more than 90 percent of the subjects were Black.

There was a statistically significant relationship between childhood lead exposure and the total number of arrests—more lead was associated with more arrests. The relationship with childhood lead exposure was even stronger when researchers only looked at arrests for violent crimes. The kids who had high levels of lead in their blood at age six were, on average, arrested almost 50 percent more often than the kids who had low blood lead levels at age six. The group of six-year-olds with the *highest* blood lead levels in this study averaged 18µg/dL, which is close to the *average* level for all American kids born a decade earlier, around 1970. The authors of this study carried out a statistical analysis suggesting that for every million kids poisoned with that level of lead, we might expect 390,000 more arrests, including 87,000 extra arrests for violent crimes, every year.[14]

One of the most extensive studies ever carried out of childhood lead exposure and later criminal activity at the individual level was published in Sweden in 2017.[15] The researchers were able to compare rates of childhood lead exposure and later criminal convictions for about 800,000 individuals. This impressive data set was available because Sweden has excellent record keeping, and also a long-standing program of monitoring air pollution using moss. Yes, moss. Mosses (bryophytes) are tiny plants that grow on rocks and other surfaces, and because they have no roots, they acquire all of their water and nutrients through the air. So, any heavy metals (such as lead) that are in the air in a particular location will also end up in the mosses.

The Swedish Environmental Protection Agency has been collecting moss samples from 1,000 locations around the country every five years since 1975. Scientists measure the amount of lead and other toxins in the new growth that the moss has produced during the preceding three years. Numerous studies have demonstrated that the amount of lead in the moss is an accurate indicator of the amount of lead in the air during those three years. The authors of this study focused on individuals born in Sweden between 1972 and 1984, a period during which the amount of lead in Swedish gasoline dropped dramatically. Because Sweden had already banned the use of lead paint back in the 1920s, airborne lead from leaded gasoline was the primary route of exposure for these children. Previous studies have demonstrated a tight correlation between airborne lead levels, as measured through moss sampling, and blood lead levels in Swedish children. Those moss samples give us a pretty clear picture of how much lead was in the kids in a particular neighborhood within any particular three-year period in the 1970s and 1980s.

During the phaseout of leaded gasoline in Sweden, lead levels changed differently over time in different neighborhoods, creating a useful natural experiment for examining the long-term effects of childhood lead exposure. Sorting individuals by both the year and location of their birth allowed these researchers to control for many factors (neighborhood differences, changing economic circumstances, etc.) that can influence criminal behavior. The researchers examined the amount of airborne lead that each individual was exposed to during ages one to three, along with records showing whether that individual was convicted of a crime by the time they turned twenty-four. Using convictions as a measure of criminal activity has drawbacks due to biases in the criminal justice system; however, racial differences do not play a substantial role in this study because Sweden is less racially diverse than the United States.

The researchers found that the kids who were exposed to more lead as children were significantly more likely to be convicted of a crime later on. The impact of lead exposure on criminal convictions held up even when the researchers controlled for various individual, family, and neighborhood characteristics. This relationship between childhood lead exposure and later criminal activity was stronger for boys than for girls, and was statistically significant for all blood lead levels above 5μg/dL. This study of 800,000 individuals, born in different neighborhoods in different years, provides strong evidence of the link between childhood lead exposure and later criminal activity.[16]

Dr. Needleman's study in Pittsburgh, the two Cincinnati studies, and the 2017 Swedish study all looked at individuals. Many other studies have examined the relationship between rates of childhood lead exposure and later rates of violent crime at the population level, on a local, national, or international scale. We've already talked about why our definitions of "crime" and "violent crime" are problematic. When we try to measure the *rate* of crime, or violent crime, in different locations, at different points in time, things get even trickier. Before we start looking at the correlations between rates of childhood lead exposure and rates of violent crime, let's take a moment to consider some of the challenges in accurately measuring crime rates.

There are two ways to estimate the rates of different types of crime, and they can sometimes give you different results. One method is to ask people whether or not they've been the victim of a crime. In the United States, this is done through the National Crime Victimization Survey (NCVS). Every year, employees of the U.S. Department of Justice interview members of around 135,000 households. The households are chosen through a complicated randomization process designed to ensure that they are representative of the country as a whole. Interviewers then ask the members of these households a set of very specific questions about different types of crime that could have happened to them.[17]

Clearly, there are some problems with this method. Some people may be reluctant to share their experiences with a government employee. As with any survey that involves self-reporting, people may either intentionally or inadvertently give false information or leave things out. However, there are also advantages to using the National Crime Victimization Survey to estimate crime rates. For example, the list of questions is extensive and specific, asking about particular types of incidents. Also, a victimization survey includes crimes that weren't reported to the police as well as crimes that were reported.

The other method that researchers use to estimate crime rates is by adding up all the crimes reported to law enforcement. The Uniform Crime Reports (UCR) database put together by the Federal Bureau of Investigation compiles information that is reported to the FBI by almost 18,000 law enforcement agencies, including federal, state, county, and city agencies, along with colleges and universities and American Indian tribes.[18] Each of these organizations sends in a monthly report, including all of the crimes that were reported to them, or that police or investigators found out about in some other way.

The big problem with using Uniform Crime Reports data is that not all crimes get reported. On one hand, some communities have better relationships with law enforcement than others, making them more likely to report crimes that occur, and some types of crime are much more likely to be reported than others. In addition, the accuracy of the database depends on all of the thousands of different law enforcement agencies who contribute to it. On the other hand, there are advantages to using the Uniform Crime Reports database. The UCR covers a wider variety of types of crime than the NCVS, and the UCR's focus on *reported* crimes overcomes some of the problems of using survey data, including the difficulty of remembering past events accurately. The biggest advantage of the Uniform Crime Reports database is that it goes all the way back to 1929, while the NCVS didn't start until 1972. Plus, the NCVS underwent a significant redesign in 1992, so the first twenty years of results are known to be less accurate than the more recent data.

When we look at the nationwide rate of violent crime since the mid-twentieth century using Uniform Crime Reports data, we see two striking trends. The first is this: violent crime rates rose throughout the 1960s, 1970s, and 1980s, and peaked in the early 1990s. When I talk to people about the history of violent crime in the United States, they usually aren't surprised by this first trend. Many people still remember the infamous "Willie Horton" ad that tanked the presidential campaign of Michael Dukakis in 1988 by painting him as "soft on crime." Crime was a big issue, especially urban crime, as depicted in movies like *Do the Right Thing* (1989) and *Boyz n the Hood* (1991). When I was in high school, some friends and I wrote a play called *Alice in New York*, in which an increasingly alarmed Alice meets a brothel owner, a drug dealer, and a couple of guys selling stolen watches. We had good reason to imagine urban life as full of violence and danger—by the time we started our senior year of high school in 1993, violent crime rates throughout the United States, and especially in urban areas, had been rising for three decades.

However, the Uniform Crime Reports database also shows a second trend that many people are still unaware of: from the early 1990s to the mid-2010s, violent crime rates in the United States dropped precipitously. In fact, the overall rate of violent crime now is only about half of what it was in the early 1990s. That means that a person living in the United States this year is only about half as likely to be the victim of a violent crime as a person living in this country in 1993 was, according to UCR data. The crime drop since the early 1990s was even steeper in the National Crime Victimization Survey data. Criminologists have written volumes about the "crime drop," but the average American doesn't seem to know about it. In Gallup public opinion polls since 1993, typically more than 60 percent of respondents believe that crime rates are rising, while only around 20 percent believe that crime rates are declining. Even during the steepest decline, in the late 1990s, a majority of Americans told pollsters that they thought national crime rates were rising.[19]

If we live in a safer country now than we did a quarter century ago, why have so many Americans failed to absorb this good news? Research has shown that most of what people know about crime comes from the media,[20] and now that we're exposed to more and more media throughout the day, we're exposed to more and more coverage of violence. Many of us also have a sense that rising crime rates are normal—that increasing violence is a natural result of the alienation of modern life. However, as Dr. Steven Pinker has exhaustively demonstrated in his 832-page book *The Better Angels of Our Nature: Why Violence Has Declined*, rising levels of violence are far from the norm.[21]

Dr. Pinker's research demonstrates an overall global trend over the past few millennia toward less violence, not more. If violence in general tends to lessen over time, then the recent decline in violent crime in the United States isn't as surprising as the unusual rise in violent crime during the middle part of the twentieth century. Undoubtedly, a number of factors—social, economic, technological, and so on—contributed to this crime wave. One of those factors, according to a growing body of scientific research, was the rising use of leaded gasoline and the resulting increase in lead exposure among American children.

In 2000, economist Rick Nevin published a paper that included a striking graph.[22] First, Nevin plotted average exposure to gasoline lead over time, based on U.S. Geological Survey data about the amount of lead used in gasoline nationwide, and the total U.S. population. As we've seen, the result is a hump-shaped curve, with exposure to gasoline lead rising throughout the 1940s, 1950s, and 1960s, peaking in the early 1970s, and then declining dramatically during the late 1970s and throughout the 1980s. Next, Nevin plotted the overall rate

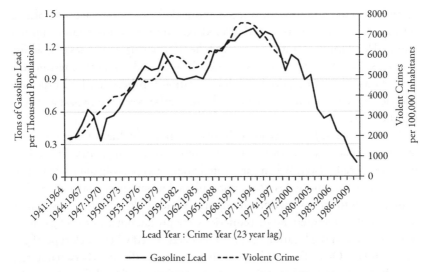

FIG. 6.1 The relationship between gasoline lead exposure and violent crime rate, with a twenty-three-year lag, from Rick Nevin, "How Lead Exposure Relates to Temporal Changes in IQ, Violent Crime, and Unwed Pregnancy," *Environmental Research* 83, no. 1 (2000).

of violent crime over time, based on the Uniform Crime Reports database. Again, we see a hump-shaped curve, with crime rates rising in the 1960s, 1970s, and 1980s, peaking in the early 1990s, and then declining sharply after that. Finally, Nevin shifted the lead exposure curve by twenty-three years, representing the time required for lead-exposed babies and toddlers to grow up into crime-committing young adults. The two curves match almost exactly, as you can see in Figure 6.1.[23]

Of course, correlation is not causation. Writing in *Mother Jones*, Kevin Drum pointed out that sales of vinyl LPs also rose throughout the 1940s, 1950s, and 1960s, then declined in the 1980s and 1990s.[24] Yet, nobody is blaming childhood exposure to vinyl records for later crime rates. Meaningless correlations are everywhere. Tyler Vigen has compiled a whole collection of them in his book *Spurious Correlations*,[25] in which we learn that, for more than a decade, the number of people who died by falling into a swimming pool each year was correlated with the number of films that Nicholas Cage appeared in. The divorce rate in Maine is quite closely correlated with the rate of per capita margarine consumption. Just because two curves look the same doesn't mean they're actually related to each other.

Nevin's argument wasn't simply based on correlation. We know about the changes that occur in developing brains when they are exposed to lead, and how those same types of changes are related to the odds of committing a violent crime. As we saw in the studies described above, individuals who were exposed to more lead in childhood are, indeed, more likely to commit acts of violence during adolescence and young adulthood. Nevin showed this relationship at a national level for the United States, and the relationship continues to hold for the years since his 2000 paper was published. In addition, other researchers have shown this same population-level relationship between lead exposure and violent crime, with a similar two-decade lag between the two, at a variety of other scales.

Economist Dr. Jessica Wolpaw Reyes hit upon a brilliant way to test the hypothesis that changes in lead exposure contribute to changes in violent crime rates two decades later. When the United States was phasing out leaded gasoline in the 1970s and 1980s, some states phased it out faster than others. The main reason for this variation among states had to do with how gasoline is refined and distributed. Some refineries were producing lower-lead gasoline than others, and the locations of pipelines determined where each type of gas would end up. Other factors also contributed to the differences between states, including the types and ages of cars (which are impacted by differences in climate) and the average number of pumps available at each gas station. Overall, the differences in the lead content of gasoline in each state during the phase-out years were, according to Dr. Reyes, "substantial and largely random."[26]

Dr. Reyes gathered information on violent crime rates over time for each state from the Uniform Crime Reports database. When she compared the amount of gasoline lead in each state with the amount of violent crime in that same state two decades later, she found a consistent pattern. In states that got the lead out of their gasoline more quickly, violent crime rates dropped more quickly after the two-decade lag. The states that were slower to get the lead out of their gasoline eventually experienced a slower drop in violent crime. These results held up even when Dr. Reyes used statistical techniques to account for the effects of people moving from one state to another, and when she subtracted out the effects of a long list of other factors that have been shown to influence crime rates.[27]

More recently, toxicologist Dr. Howard Mielke (whose earlier research was discussed in chapter 3) and economist Dr. Sammy Zahran looked at the relationship between childhood lead exposure and later violent crime rates in six different cities around the United States.[28] They used gasoline lead data and traffic information to estimate the amount of lead released into the air from vehicles in each city in each year. They then compared those estimates with the number of reported aggravated assaults in each city two decades later. In each city, the rise and fall of gasoline lead predicted the later rise and fall in assault rates.

Even in the different neighborhoods of a single city, rates of lead exposure are associated with later rates of violent crime. A group of researchers in St. Louis looked at different census tracts within the city.[29] A census tract typically contains a few thousand people, so we can think of these as different neighborhoods. The researchers looked at the blood lead levels of almost 60,000 preschool-age children in different neighborhoods and found that the neighborhoods with the highest blood lead levels would go on to experience the highest rates of violent crime in the years that followed. As with other studies we've seen, this relationship between childhood lead exposure and later violent crime rates held up even when the authors subtracted out the effects of socioeconomic differences between neighborhoods.

It's not just the United States—this relationship between lead and violent crime shows up in country after country. In 2007, Rick Nevin published a study of nine countries around the world.[30] Nevin estimated childhood lead exposure in each country for each year using both leaded gasoline data and actual blood lead measurements. He also gathered crime statistics from law enforcement agencies in each country. All nine countries experienced the same rise and fall in gasoline lead, but the timing was different—some nations got the lead out of their gas earlier than others. Nevin found that, in every country, changes in childhood lead exposure predicted changes in crime rates two decades later.

Back in the 1960s, the United States had some of the highest rates of lead exposure in the world. It is estimated that by 1960, three-quarters of all the leaded gas ever used in the world had been used in the United States,[31] and the United States continued to burn leaded gas at increasingly higher rates throughout the following decade. So American kids in the 1960s and early 1970s were exposed to far more gasoline lead than kids in most other countries. Then, in the 1980s and early 1990s, the United States had higher rates of violence than any other developed country. Although many European countries had similar rates of nonviolent crimes like burglary, by the mid-1990s the U.S. rate of "death and life-threatening injury from intentional attacks" was four to eighteen times higher than in other developed nations.[32]

The United States started phasing out leaded gasoline in the 1970s, while most European countries didn't begin regulating gasoline lead until the 1980s.[33] So, while rates of violent crime in the United States began dropping by the mid-1990s, the countries that were slow in switching to unleaded gas generally had their "crime drop" later than those of us who made the switch earlier. Great Britain, for example, phased out leaded gasoline starting in the mid-1980s, and their rate of violent crime peaked in the mid-2000s and then dropped rapidly, just as Nevin had predicted.[34]

It is important to note that rates of violent crime in other developed countries were never as high as they were in the United States in the 1980s and early 1990s. Criminologists are still debating the many interacting factors that led to this uniquely American violent crime wave. Certainly, access to firearms was a major factor that set us apart from many of our fellow nations.[35] The crack epidemic, as well as a number of social and economic factors, played a role as well. Childhood lead exposure alone cannot explain all of the differences in violent crime rates between the United States and other developed nations, but the consistent relationship in timing between changes in childhood lead exposure and later changes in crime rates in various countries suggests that lead was, indeed, one of the many factors influencing the rise and fall in rates of violent crime.

The link between childhood lead exposure and later violent crime rate shows up on the local scale outside of the United States as well. In 2016, a group of researchers in Australia studied the relationship between airborne lead emissions and rates of violent crime approximately twenty years later.[36] When they looked at different states, and even at individual suburbs, they found a strong relationship between lead exposure and later crime rates, even when they subtracted out the effects of a variety of sociodemographic factors.

So, various researchers have found a consistent relationship between childhood lead exposure and later crime rates at the international, national, state, and city scale. This relationship is exactly what we would expect from the many, many studies, on both animals and humans, showing the negative impacts of lead on brains and behavior. Nevertheless, there is still significant controversy surrounding this research. One issue that critics have raised is the possibility that there was some other factor that happened to rise and fall along with leaded gasoline but is not accounted for in the models these researchers have used. However, the evidence of a relationship between lead exposure and crime rates is not limited to studies of leaded gasoline.

Back in 1998, a group at Dartmouth College estimated differential lead exposure for every county in the United States based on the locations of industrial facilities that release airborne lead.[37] Although their work did not include a time lag to account for lead-exposed children growing up to become crime-committing young adults, they point out that the locations of lead-polluting industrial facilities don't change very rapidly over time. The counties that had lead smelters in the 1990s were mostly the same counties that had lead smelters in the 1970s. After controlling for ten socioeconomic and demographic variables, the researchers found a significant relationship between rates of exposure to industrial lead and rates of violent crime.

In 2016, Dr. James Feigenbaum from Harvard and Dr. Christopher Muller from Berkeley found another way to examine the impact of lead exposure on crime rates.[38] Many cities across the United States installed public water systems in the late 1800s. Some of these city water systems used iron service pipes, and some used lead service pipes. The decision about which metal to use was mainly based on transportation costs—cities that had one or more lead smelters nearby typically chose lead pipes, and cities far from the nearest lead smelter typically chose iron. (Lead is very heavy, so it's expensive to transport over long distances.) Dr. Feigenbaum and Dr. Muller gathered data about which type of pipes was installed in each city, and also about how acidic the drinking water is in each city. Over time, acidic water dissolves the lead pipes, carrying lead into the homes and businesses using the water. The lead pipes stay mostly intact if the water isn't acidic, so there's not as much lead exposure for the people drinking that water.

Then Dr. Feigenbaum and Dr. Muller looked at homicide rates in these same cities in the 1920s and early 1930s. They chose to study homicide because there are relatively reliable records available all the way back to the early 1920s. They found that the high-lead-exposure cities had significantly higher murder rates than the low-lead-exposure cities. As with the other studies we've looked at, the relationship between lead exposure and murder held up even when the researchers accounted for a long list of other factors that might influence murder rates. This study and the Dartmouth study of industrial lead facilities demonstrate that the rise and fall of leaded gasoline is not the only example of increased lead exposure being associated with higher rates of violent crime.[39]

Another criticism of the national-, state-, and city-level analysis of lead exposure and violent crime rates is their use of the Uniform Crime Reports database as the source of information about violent crime rates. In 2016, criminologists Dr. Janet Lauritsen, Dr. Maribeth Rezey, and Dr. Karen Heimer analyzed differences in violent crime trends in the UCR database and the National Crime Victimization Survey for the period from 1973 (when NCVS started) to 2012.[40] Although both data sources show the same trends from the mid-1980s forward, they differ before that. The UCR shows sharply rising violent crime rates in the 1970s and 1980s, while the NCVS shows violent crime rates being basically flat during that time. Dr. Lauritsen, Dr. Rezey, and Dr. Heimer argue that the NCVS violent crime rates are more reliable, because they correlate better with homicide rates, and we know that data about homicides are quite reliable.

The researchers suggest that the reliance on UCR data undermines the connection between childhood lead exposure and violent crime rates.[41] During the period from 1973 to 1990, the National Crime Victimization Survey actually recorded a slight reduction in violent crime. This is very different from the crime spike we see during this time in the Uniform Crime Reports database, which correlated with rising lead exposure two decades earlier. However, the Bureau of Justice Statistics itself has stated that its National Crime Victimization Survey undercounted some crimes before the survey was redesigned in 1992.[42] (Dr. Lauritsen, Dr. Rezey, and Dr. Heimer used a statistical technique to adjust the pre-1992 data.) Even if the three criminologists are right, and violent crime rates weren't continually rising throughout the 1970s, 1980s, and early 1990s, the NCVS still shows high rates of violent crime during that time, when the most lead-poisoned kids were teenagers and young adults. Then both the NCVS and the UCR show steep declines in violent crime after the early

1990s, as kids who grew up around unleaded gas entered their prime crime-committing years. In fact, the "crime drop" is even steeper in the National Crime Victimization Survey. No matter how you look at it, kids who were less lead-poisoned committed fewer violent crimes.

Another area of controversy is the question of just *how much* of an impact childhood lead exposure has had on violent crime rates. Some researchers have claimed that changing childhood lead exposure is responsible for most of the change in violent crime rates in the past half century, and critics have taken exception to that claim. Many have rightly highlighted the various other factors that have been shown to contribute to the rise and fall of violent crime. Crime is affected by countless factors, and those factors can interact with each other in complicated ways. The argument that childhood lead exposure affects violent crime rates does not undermine the importance of any of the multitude of other factors that we know can affect crime rates. In fact, the neuroscience research suggests that exposure to lead in childhood makes people even more susceptible to some of these other factors.

For example, multiple studies have found that lead exposure has a stronger negative impact on children of low socioeconomic status.[43] Lead exposure and chronic stress have been shown to harm the same parts of the brain. One study in rats showed that the presence of lead magnified the amount of damage that stress does to developing brains.[44] Another rat study found that the consequences of early lead exposure were reduced if the baby rats were raised in a "stimulating" environment.[45] All of these studies suggest that exposure to lead can exacerbate the negative impacts of a stressful or neglectful childhood environment. In addition, lead has been shown to harm the cells responsible for repair of the brain after an injury.[46] Studies have shown that having a traumatic brain injury as a child increases the chances of criminal behavior later on,[47] so if lead reduces a person's ability to recover after such an injury, then lead exposure would exacerbate the effect of brain trauma.

Recognizing that childhood lead exposure affects criminal behavior does not diminish the importance of childhood stressors or brain injuries; it enhances their importance. The same is true for many other factors that contribute to violent crime. We would expect a lead-injured brain to be particularly impaired by substance abuse, less able to cope with family and social dysfunction, and less able to succeed in a subpar educational system. The factors that influence rates of violent crime can't be divvied up like wedges on a pie chart, because each one may intensify or diminish some of the others.

Criminologists Zimring and Hawkins argue that "there is no reason why diverse factors ranging from television programming to loaded guns cannot play different types of causal roles in violence."[48] They also point out that "two separate exacerbating problems can more than double the negative impact they cause jointly."[49] Zimring and Hawkins call the tendency to focus on a single explanation for shifting violent crime rates, to the exclusion of other potentially important factors, "probably the most frequently found error in the rhetoric of violence."[50] Clearly, childhood lead exposure isn't the only explanation for the dramatic rise and fall of violent crime rates in the past half century, but a substantial body of research suggests that it was one of the factors involved.

We know that lead harms the developing brain; we can see this at the cellular level in animals and in the larger structures of the human brain through advanced brain scanning technology. From both animal studies and human studies, we know that these changes in the brain cause problems with learning, attention, and impulse control. Even when we subtract out the effects of numerous other demographic, economic, and social variables, we can see the impact of childhood lead exposure on crime and violence at the individual level and the population level. Counties with lead industries have more crime, and cities with high-lead-exposure public water systems historically had more murders. Although the use of leaded gasoline rose and fell at different times in different countries, states, and cities, we consistently see a rise and fall in violent crime rates that takes place about two decades later.

All of these studies, taken together, make a strong case that the effect of childhood lead exposure on violent crime rates over the past half century has been substantial. There were clearly decades when this country was more dangerous than it would have been if we'd kept the lead out of our gasoline back in the 1920s. We are undoubtedly safer now than we would be if we hadn't taken the lead out of our gasoline in the 1970s and 1980s. Yet, since I started working on this book, I have talked to a number of highly educated, well-informed people who have never heard of the link between lead and crime. If the evidence linking childhood lead exposure to violent crime is so overwhelming, why don't more people know about it?

One reason is this: lead exposure is studied by natural scientists, and crime is studied by criminologists. These two groups of experts typically have very different training, which leads them to think about the world in very different ways. At many universities—including mine—natural scientists and social scientists work in different buildings, on different parts of the campus. A toxicologist and a criminologist could both work for the same institution for their entire careers and never say more than a few words to each other at a faculty meeting. They would publish in different journals and attend different conferences. There is no reason why they would necessarily know about each other's work. The link between lead and crime is a great example of why interdisciplinary studies are so important. Too often, researchers who venture beyond their own narrow area of expertise are not considered to be doing "serious" work, and are punished professionally.[51]

Another reason why the link between lead and crime is not more widely discussed is that researchers can fall victim to an all-too-human tendency to

defend their own theories and reject everybody else's. Studies examining policing, incarceration, family structure, neighborhood organizations, drug use,
and even abortion laws, among other factors, have found evidence that these
factors have substantially influenced crime rates over time.[52] It is easy to make
the mistake of thinking that if one explanation is more important, that must
make other explanations less important. Yet, as we have seen, the various factors
may influence one another in numerous, complicated ways, so highlighting the
importance of childhood lead exposure does not diminish the importance of
these other factors. Openness to considering new factors is especially important because lead isn't the only toxin in the environment that has been shown
to alter human behavior. A number of other toxic metals and organic chemicals can also alter the brain and change people's learning and behavior.[53] Additional interdisciplinary research is needed in order to understand the
consequences of these toxins for individual humans and for society at large.

In addition to disciplinary obstacles to studying the lead/crime link, there
are political objections as well. Some on the right would prefer to focus on the
social pathologies that are associated with increased crime rates, and on moral
and/or religious solutions to those problems.[54] If we highlight lead exposure as
a contributor to crime, then one solution is government intervention to reduce
lead in the environment, and some on the right see environmental regulations
as an unacceptable expansion of government power.[55] On the other hand, if
crime can be attributed to the breakdown of the nuclear family, or to secular-
ization, or to violent movies and video games, then solutions would include the
promotion of marriage, prayer in school, and a return to more "wholesome"
media. If you already have the solutions in mind, then it makes sense to focus
on the problems that would necessitate those solutions.

There are also objections on the left to any discussion of the lead/crime link.
For one thing, biological explanations of criminality have an incredibly sordid
history.[56] Many early criminologists believed that a propensity to crime was
genetic, and could be identified by the shape of a person's forehead and jaw or
other physical characteristics. A belief in the inheritability of criminality has
been used to support racist policies in the past. Given the historical and current racial disparities in childhood lead exposure, it is important that a focus
on lead exposure not become a distraction from the very real racial biases that
continue to exist in our policing and criminal justice policies, even as racial disparity in crime rates has shrunk considerably.[57]

We know that police practices demonstrate racial bias. Stop-and-frisk policies have been disproportionately used against Black and Latino men.[58] Black
drivers are more likely than White drivers to be pulled over, to be ticketed, and
to have their cars searched,[59] and one study found that Black Americans are
2.5 times more likely to be killed by police officers than White Americans.[60]
These disparities in policing cannot be wholly attributed to differences in rates

of criminal behavior. For example, although Black and White Americans are about equally likely to use marijuana, Black people are almost four times more likely than White people to be arrested for marijuana possession.[61] These racial disparities exist throughout the criminal justice system. Studies have shown that prosecutors tend to seek harsher penalties when the defendant is Black and/or when the victim is White.[62] Mass incarceration has so disproportionately affected Black communities that it is being called "the new Jim Crow."[63]

The existence of racial bias in every part of the system—police, prosecutors, judges, juries, correctional and parole officers—has been shown in study after study.[64] In some cases, these outcomes result from the unconscious racial bias that exists in almost everybody.[65] For example, one 2008 study used a word-association test to examine unconscious bias in a group of trial judges.[66] It found that the judges did have measurable racial bias, and that this bias affected their decisions in a series of hypothetical cases. The prevalence of both conscious and unconscious bias means that people of color are treated more harshly than White people in their interactions with police officers, courts, and correctional officers, leading to racial differences in incarceration rates that cannot be explained by differences in criminal behavior. Anybody who tries to use the link between lead and crime to explain away the very real racism of our criminal justice system is misrepresenting the data.

Another possible misuse of information about the lead/crime link relates to risk assessment algorithms that are of great concern to many civil liberties advocates. These computer programs are increasingly being used throughout the criminal justice system, affecting bail amounts, sentencing, and parole decisions.[67] The programs use information about individuals to rate them in terms of their predicted risk of committing a crime in the future. Shockingly, some states are using systems designed by for-profit companies, based on underlying algorithms that are not available for public scrutiny.[68] People are being given higher bail amounts, longer prison sentences, and delayed release from prison because of an undisclosed mathematical analysis of their personal history. As you might expect, studies have found these predictions to be racially biased.[69] Adding lead exposure information to these flawed algorithms wouldn't make them any more effective, and would introduce yet another source of racial bias.

Once we look past our academic and political blinders and see the importance of childhood lead exposure as an influence on violent crime rates, we can see some important ramifications for society. One implication is that previous lead exposure could be considered as a mitigating factor in the criminal justice system. Being exposed to lead doesn't force anybody to commit a violent crime; however, our legal system has a tradition of taking a defendant's personal history into consideration during sentencing. A person may be considered less culpable in light of a difficult childhood or other challenging circumstances. Certainly, it makes legal sense to consider childhood lead exposure in this category. It makes philosophical sense as well. We all have violent impulses from time to time. ("I want to strangle that guy.") The research suggests that the damage to the brain caused by lead makes it more difficult for a person to control those violent impulses. So, even as we hold each person responsible for controlling their own violent impulses, we can recognize that this moral responsibility is easier for some of us than others.

At the same time that we hold individuals responsible for the violent acts that they commit, we can also hold the lead industry and the oil companies responsible for their contribution to the rising crime rates of the 1960s, 1970s, and 1980s. Just as the causes of a particular effect (like violent crime rates) cannot always be apportioned like wedges in a pie chart, so I would argue that moral responsibility is not a zero-sum game. Every ton of lead these industries were responsible for releasing into the atmosphere contributed to an increase in the blood lead levels of American children, and rising blood lead levels increased the chances of those children going on to commit violent crimes as adolescents and adults. The producers of that lead bear some responsibility for those violent acts. Conversely, the scientists, activists, and government officials who forced the oil companies to switch to unleaded gasoline deserve some of the credit for the subsequent drop in crime rates. To some degree, we all owe our safety to the brave people who fought to get the lead out of our gas tanks.

The lead-poisoned members of Generation X (and, to a lesser extent, other generations) who did commit violent crimes were themselves the victims of a type of violence. The release of lead into our communities represented a type of structural-level violence that affected all of us. But it is important to remember that the violence of lead poisoning was not perpetrated on all communities equally. Poor communities, inner-city communities, and Black communities were much more poisoned, on average, than affluent, rural/suburban, or majority-White communities.

Companies continued to put lead into gasoline—and paint, plumbing fixtures, and so on—long after they had been shown the harm that lead could do. The free market couldn't protect Americans from that harm—it took a long, hard fight to secure the federal regulations that got the lead out of gasoline. In the end, it was a fight that all Americans won, because the switch to unleaded gas contributed to an unprecedented quarter-century drop in violent crime rates. Back in the early 1990s, when crime rates had been rising for decades, many sociologists and criminologists sounded the alarm, arguing that demographic trends would continue to push violent crime rates higher for at least another decade.[70] In fact, according to the UCR database, violent crime fell by about 50 percent from the peak in the early 1990s to the mid-2010s.[71] The NCVS shows a drop in violent crime rates of almost 75 percent during that time.[72] Although the past few years have seen a leveling off in crime rates, or even a small increase, depending on which data set we look at, the rate of violence is still far, far below what it was twenty-five years ago.

During the 1980s and early 1990s, experts were especially worried about the rise in youth violence. In 1996, criminologist James Fox submitted a report to the U.S. Bureau of Justice Statistics in which he argued that "we likely face a future wave of youth violence that will be even worse than that of the past ten years."[73] In fact, youth violence had already begun to decline by then. As overall rates of violent crime have fallen in the past quarter century, the decline in violence committed by teenagers has fallen the most dramatically. In the early 1990s, juveniles were about twice as likely to be arrested for homicide as adults in the United States, but now the probability of being arrested for murder is about the same for juveniles and adults.[74] Overall, arrest rates for those under the age of forty have declined substantially since the early 1990s, while arrest rates among people over age forty (a category that now includes members of more lead-poisoned generations) have actually increased modestly.

The overall decline in violence has occurred all across the United States, especially in our biggest cities. During the crime wave of the early 1990s, large cities were especially dangerous places to be. On average, large cities had violent crime rates that were almost double those of smaller cities, and almost three times higher than rural areas.[75] Many factors contributed to this disparity in crime rates, and childhood lead exposure was one of them. Large cities have a higher concentration of traffic than smaller cities or rural areas, so when cars and trucks were burning leaded gasoline, the largest amounts of lead were being released in inner-city neighborhoods.

As crime rates came down, they declined most precipitously in big cities. The difference in violent crime rates between larger and smaller cities has been shrinking for the past quarter century. A 2013 study showed that if you include accidental deaths as well as homicides, cities are actually significantly safer places to live than rural areas.[76] New York City is an especially vivid example of the decline in violent crime in large cities. In 1990, there were 2,245 murders in New York City. By 2014, that number was down to 328, even though the overall population of the city had increased. The murder rate had declined by 87 percent.[77] Criminologist Franklin Zimring has pointed out that this precipitous decline in violence in New York City occurred despite no significant changes in "populations, schools, transportation, and economy."[78] Since the unemployment rate in New York City was actually slightly higher in 2014 than in 1990, economics alone cannot account for this dramatic reduction in the murder rate.[79] Overall, large cities are now much safer places to travel, work, and live than they were a quarter century ago.

The whole country is much safer than it used to be, but based on public opinion polls, it seems that Americans are not generally aware of this fact, including American parents. Several studies have found that a fear of crime contributes to parents putting more constraints on their children's activities.[80] The current generation of American children spends less time outdoors, and less time without direct adult supervision, than previous generations did,[81] even though some of those previous generations of children were growing up in more dangerous times. There is now an effort under way to reverse these trends and give kids more independence. In her *New York Times* best seller, *How to Raise an Adult*, Dr. Julie Lythcott-Haims, the former dean of freshmen at Stanford University, describes how a lack of independence can lead to depression and anxiety.[82] A growing movement called "free range parenting" holds that children should have much more freedom to go places and do things unsupervised. A mom named Lenore Skenazy started the blog *Free Range Kids* after being called "the worst mom in America."[83] Her offense? She let her mature, savvy nine-year-old son ride the New York City subway alone. (He got home just fine.) Many of her detractors are unaware that riding the subway is now much, much safer than it was when they themselves were kids.

We have many people to thank for the safer cities and towns we live in today. Community organizations, public officials, and researchers across many disciplines played important roles in reducing the rate of violent crime. So did everyone involved in the elimination of leaded gasoline. The precipitous decline in childhood lead exposure in the United States in the past half century, as well as the steep decline in violent crime rates in the past quarter century, has provided an abundance of positive consequences. Unfortunately, not all the news is good. Violent crimes still happen every day, of course. And far too many children are still being exposed to harmful levels of lead.

7

The Lead
Problem Persists

━ ━ ━ ━ ━ ━ ━ ━ ━

Getting the lead out of our gasoline was a major victory in the fight against childhood lead exposure. There was a dramatic decline in the average amount of lead in the blood of American preschool children between the mid-1970s and the mid-1980s, and the evidence suggests that most of that drop can be attributed to the phaseout of leaded gas. There have been other victories as well. Lead paint, lead solder for food cans, lead pipes, and leaded plumbing fixtures have all been outlawed. The average blood lead level of small children in the United States continues to go down every year. We've come a long way.

However, the problem of childhood lead exposure is far from solved. We now know a lot more about the harm that even very low levels of lead can do to the developing brains of children. Once, we would have considered a child with a blood lead level under 25μg/dL to be "safe." As scientists carried out more studies on the effects of low-level lead exposure, this cutoff was reduced to 10μg/dL and then 5μg/dL, but research shows that there is no safe level of lead for children. Many have argued that a blood lead level of 2μg/dL should be considered "elevated," and efforts should be made to reduce exposure to lead in all kids above that level.[1] By that standard, we have a long way to go in protecting American children from lead.

The issue of childhood lead exposure burst into the national media in 2015 with revelations about lead-contaminated drinking water in Flint, Michigan. Flint was once a thriving manufacturing town—General Motors was founded there in 1908, and the town's nickname was "Vehicle City." Like many American manufacturing towns, Flint has gone through hard times in recent decades. General Motors slashed its workforce there, and other manufacturing jobs have been lost as well, leading to a declining and increasingly impoverished population. Flint's hard times were chronicled in the 1989 documentary *Roger & Me* by Michael Moore, which depicted one down-on-her-luck Flint resident selling rabbits "for pets or meat." By the early 2000s, the city was deeply in debt. In 2011, the state of Michigan took over, appointing an emergency manager to run Flint.[2] The governor chose a White man to manage this majority-Black city.

For half a century, Flint had purchased its city drinking water from Detroit, which takes its water from Lake Huron. Flint, like many cities throughout the United States, has lead pipes as part of its water-delivery system. As we saw in chapter 6, the amount of lead in drinking water depends on what type of pipes the water is running through, but also on characteristics of the water itself. More acidic water tends to dissolve the lead in the pipes, in a process called corrosion. The lead that dissolves into the water then ends up coming out of the faucets of consumers. "Corrosion control" involves adjusting the pH of the water to make it less acidic, and also adding chemicals that safely coat the inside of the pipes, creating a barrier that prevents the lead from dissolving into the water. Careful corrosion control can often keep the level of lead in the drinking water quite low, even if it is flowing through lead pipes. As long as Flint was getting its water from Detroit, these corrosion control procedures were safely in place.[3]

One of the cost-saving measures that the state-appointed emergency manager of the city of Flint implemented was a change in the city's water supply. The plan was to build their own pipeline to Lake Huron eventually, but in the meantime, the city switched to using water from the nearby Flint River, starting in April 2014. The Flint River water was highly corrosive, and city water managers failed to implement proper corrosion control. The protective layer that had built up inside the city's old lead pipes was eaten away, and lead started to dissolve into the drinking water of the city's residents. Water from the Flint River also contained high levels of various toxins and bacteria. Soon, people living in Flint noticed changes in the color, taste, and smell of their drinking water, and began reporting health problems such as rashes and hair loss. Still, the city continued to use Flint River water.[4]

In February 2015, a water expert at the Environmental Protection Agency (EPA) named Miguel Del Toral tested the water at the home of Flint resident LeeAnne Walters and found a level of lead that was seven times higher than the EPA's acceptable limit.[5] Ms. Walters contacted civil engineer Marc Edwards, a Virginia Tech professor who had previously studied lead contamination in drinking water in Washington, D.C. Dr. Edwards and his students came to Flint and found elevated lead levels in the drinking water of 40 percent of the homes in Flint, some of them at more than one hundred times the EPA's acceptable limit.[6] Miguel Del Toral shared these findings in a June memo to his EPA superiors. Although the EPA attempted to work with city water managers to rectify the problem, the EPA did not take emergency action, and they did not notify the public. In fact, in July 2015, the month after Del Toral's memo, the Michigan Department of Environmental Quality (MDEQ) was still reassuring the public that Flint's water was safe to drink.[7]

In September, pediatrician Mona Hanna-Attisha called a press conference to announce the results of her study of blood lead levels in Flint children following the change in water source. Dr. Hanna-Attisha had found that the percentage of Flint preschoolers with blood lead levels above 5µg/dL (considered "elevated") had doubled after the switch to Flint River water. In fact, in the neighborhoods where Dr. Edwards and his team had found high levels of lead in the drinking water, Dr. Hanna-Attisha discovered that more than 10 percent of the children under age five had elevated blood lead levels.[8] Finally, in October, Flint switched back to Detroit water. Unfortunately, the damage had been done—it would take more than a year for the city's lead pipes to build up enough anticorrosion coating to make the drinking water safe again.

Sadly, Flint, Michigan, is not unique. A study reported in *Civil Engineering* magazine estimated that there are six million to seven million buildings in the United States that have lead pipes connecting them to their drinking water supply. Replacing all of these pipes will cost an estimated $35 billion to $50 billion.[9] Because of the complexity of how water quality and water treatment practices impact the amount of lead that dissolves from these pipes into the drinking water, the overall exposure of American children to lead from lead pipes is not well known. Projects are under way in Flint and other cities around the country to remove lead pipes, but at the current rate, it will take many decades before this process is complete.

The Flint water crisis highlights the complicated and often conflicting roles of government agencies in the ongoing fight against childhood lead exposure. Under the Safe Drinking Water Act, the EPA is responsible for setting the standards for safe drinking water in the United States. However, it is the individual state environmental agencies that actually monitor drinking water quality and enforce the EPA standards. Although the EPA provides guidelines for how this monitoring should be done, those guidelines are not always followed appropriately. The evidence suggests that in Flint the Michigan Department of Environmental Quality intentionally biased their sampling in order to get results that met the standards, and may have knowingly provided false information to the EPA. When the EPA's Miguel Del Toral came forward with evidence of a lead crisis, the MDEQ worked to silence and discredit him.

It seems that there were bureaucrats at the MDEQ who were more interested in avoiding conflict and safeguarding their own careers than in protecting the public from a serious health hazard. We have seen government employees engage in similarly short-sighted behavior throughout the story of leaded gasoline. For decades, the federal government allowed the sale of leaded gas, despite evidence of the harm that lead causes. Even when the Clean Air Act specifically required the agency to set standards to protect human health from airborne lead, a Reagan-era EPA dragged its feet and had to be taken to court and forced to do its job.

While government bureaucrats were among the villains of the Flint water crisis, they were among the heroes as well. EPA employee Miguel Del Toral did some of the earliest testing that helped to uncover the problem, and without the EPA, there would not be any standards limiting the amount of lead in drinking water in the United States to begin with. In the end, it is the government that has the power to set the rules that protect us from environmental harms and to enforce those rules. Without government regulations, profit-driven oil companies might still be selling leaded gasoline. Without government regulations, local water companies could set their own—probably less protective—standards for lead and other contaminants.

Government regulations play a critical role in protecting health and safety, but government employees are subject to the same perverse incentives found in many other occupations—it is usually better for one's career to cover up mistakes than to be a whistleblower. Effective environmental protection cannot be guaranteed without the work of investigative journalists uncovering malfeasance and ineptitude, independent scientists checking the work of government scientists, and legislators and the general public holding government agencies accountable.

In addition to the importance of robust government oversight, another lesson from the Flint water crisis is that environmental racism is alive and well. Around 70 percent of the residents of Flint are Black, and that percentage is even higher in the neighborhoods that were found to have the highest concentrations of lead in their drinking water. A 2017 Michigan Civil Rights Commission report identified environmental racism as an important cause of the Flint water crisis.[10] Commission cochair Arthur Horwitz explained: "We are not suggesting that those making decisions related to this crisis were racists, or meant to treat Flint any differently because it is a community of color. Rather, the response is the result of implicit bias and the history of systemic racism that was built into the foundation of Flint. The lessons of Flint are profound. While the exact situation and response that happened in Flint may never happen anywhere else, the factors that led to this crisis remain in place and will most certainly lead to other tragedies if we don't take steps to remedy them."[11]

One of the "other tragedies" that is currently being perpetuated by systemic racism in the United States is the ongoing exposure of children to lead paint. Lead paint has long been the primary source of acute lead poisoning of

American children. Lead pigments were used in paints throughout the history of the United States, and in the early twentieth century much of the house paint used in this country didn't just contain lead, it was mostly lead. Painters mixed lead oxides with vegetable oil, and applied that mixture directly to walls and trim. So, when a flake of paint chipped off the wall, that flake might have been 50 percent pure lead by weight. We know that small children will put pretty much anything in their mouths, and paint chips are no exception. Throughout the first half of the twentieth century, kids exposed to this paint were regularly suffering from acute lead poisoning, the kind that leads to seizures, coma, and even death.[12]

The amount of lead used in paints in the United States reached its peak in the 1920s, and then began to decline gradually after that as other pigments became more common. Yet Americans were still putting lead on their walls. Lead-based paint was finally banned for household use in the United States in 1978, and any housing built after that year is considered to be free of lead paint. The problem is that the United States still has a great deal of housing built before 1978, and the lead on those walls doesn't just disappear. Once a house or apartment has lead paint, it has it forever unless stringent removal techniques are used.

The focus of this book is on leaded gasoline because of its significant contribution to the dramatic rise in childhood lead exposure nationally in the 1950s and 1960s, the peak in the 1970s, and the even more dramatic drop in nationwide lead exposure during the 1980s and 1990s. The effects of the rise and fall of leaded gasoline were superimposed upon the effects of a much longer history of lead paint. No child in this country is currently exposed to airborne lead from gasoline, and no child has been for more than twenty years. The lead paint catastrophe is playing out much more slowly. Millions of American children are still potentially being exposed to unacceptable levels of lead from paint.

We now know that children don't have to eat paint chips to be exposed to harmful levels of lead.[13] (Representatives of the lead industry have falsely claimed that "the children who became lead poisoned were 'defective' to begin with, suffering from a condition called pica that led them to consume non-food items."[14]) As lead paint breaks down over time, it creates lead-containing dust. This process is sped up by the opening and closing of windows with lead-painted window frames. Lead dust can get into the air and be inhaled, and this lead dust also coats the floors and other surfaces in the home. Little kids crawl around, or toddle around, touching everything, and putting their hands in their mouths. If there is lead dust in a child's environment, there is no way to prevent that child from ingesting some of it.

As we've seen, our understanding of how much lead exposure it takes to be harmful keeps changing. The official cutoff for too much lead has fallen from 60μg/dL to 5μg/dL, and current research demonstrates that even levels below

5µg/dL can be harmful. When the most recent national survey of children's blood lead levels was completed in 2010, 2.6 percent of kids under five years old had blood lead levels of 5µg/dL or higher. This result indicates that more than half a million kids in this country still have too much lead in their blood, and most of that lead is from old lead paint.

Of course, 2.6 percent was the overall national average in 2010. In many neighborhoods, the problem is much worse. In Philadelphia, near where I live, as of 2010 more than 6 percent of little kids had elevated blood lead levels, and in 2016 (the most recent year for which data are available) the figure was still 4.4 percent.[15] Even within a single city, lead exposure is not evenly distributed. The proportion of kids with elevated blood lead levels in different zip codes in Philadelphia ranges from 0 percent to 9.8 percent. Philadelphia's Department of Public Health has found that blood lead levels within the city are closely correlated with poverty and with the proportion of the housing that was built before 1950.[16] Nationwide, Black children are more often exposed to lead paint than White children, and as a result, Black children are currently twice as likely as White children to have elevated lead levels.[17]

The challenge is that getting rid of lead paint isn't easy or cheap. Painting over it isn't good enough—if you put a fresh coat of paint on top of old paint that's disintegrating, pretty soon the new paint starts flaking off too. The only way to truly protect children is to completely remove the lead paint from their homes. Over time, experts have developed successful techniques for removing lead paint from existing homes. Specially licensed firms employ highly trained and certified individuals who can do the job, using specialized protective equipment and finishing with a thorough cleaning process. As you might imagine, this technique is very expensive.[18]

Ever since lead paint was banned in the 1970s, there have been ongoing debates about whose responsibility it is to remove all the old lead paint still on America's walls. Because lead paint is an environmental issue, a housing issue, and a health issue, the various federal agencies have passed it around like a hot potato, each one arguing that another agency should have primary responsibility. The Environmental Protection Agency runs training programs that certify contractors to do lead paint removal and "lead-safe" repairs and renovations, but it's up to property owners to actually hire these certified professionals to do the work.[19] Federal law requires that all federally owned, federally subsidized, or federally insured properties have their lead paint hazards removed, but the U.S. Department of Housing and Urban Development (HUD) has never been allocated nearly enough money to cover this enormous undertaking.[20]

The Centers for Disease Control (CDC) issued a landmark Strategic Plan for the Elimination of Childhood Lead Poisoning in 1991. The Strategic Plan recommended universal blood lead screening for all children under age five and the effective abatement of lead paint hazards in all "high-risk" housing. However, within a few years, the CDC had revised its screening recommendation from "universal" to "targeted," meaning that screening was now recommended only for children who met a complicated set of criteria. Also, it was never clear where the billions of dollars required for lead paint abatement in all of the country's high-risk housing was going to come from. The goals of the 1991 Strategic Plan have never been met.[21]

All over the country, state and municipal health departments have been working on addressing both of the goals that the CDC set forth—screening kids for lead and cleaning up old lead paint in housing where kids live. In many cities, public health workers go door-to-door in high-risk neighborhoods, offering information about the danger of lead for kids, access to blood lead measurement programs, and inspection of properties for lead paint hazards. The federal government has traditionally been an important source of funding for these on-the-ground programs, but federal funding has declined steadily for many years.[22] Here in Philadelphia, for example, the health department received $11 million in federal funding for its lead programs back in 2007, but less than $2 million in 2016,[23] and Trump administration budgets included even deeper cuts to programs that reduce children's exposure to lead paint. This dramatic decline in federal funding has led cash-strapped cities to cut their programs aimed at reducing childhood lead exposure.

Property owners have the ultimate responsibility for reducing or eliminating lead paint hazards in housing that they own. In the neighborhoods whose kids are most at risk for lead poisoning, most families don't own their own homes; they rent from landlords. By law, these landlords are required to inform potential tenants that the possibility of lead paint hazards exists in any housing built before 1978.[24] Before they sign a lease, tenants must sign a form declaring that they have been informed about this issue. Theoretically, potential tenants have the right to request a "lead hazard inspection," carried out by a certified inspector, before they sign a lease. However, demanding such an inspection may not be a practical option for low-income families desperate to find adequate housing.

If a child is exposed to lead in a rental property, the family has the option to sue the landlord (though the Black families whose children are most likely to be exposed to lead paint hazards are often distrustful of the justice system, due to their past experiences of racial bias[25]). In some cases, families have settled these suits for hundreds of thousands of dollars.[26] In other cases, landlords have prevailed in court and families have received nothing. These cases can take years to be resolved, have uncertain outcomes, and can only be initiated *after* a child has been lead-poisoned. Suing individual landlords is not likely to solve the ongoing problem of children's exposure to lead paint.

What about suing the entire paint industry? That's what the state of Rhode Island did in 1999. The state argued that lead paint constitutes a "public nuisance," and demanded that the industry responsible for this nuisance provide the funds to clean it up and protect the public from future harm. The first trial ended with a hung jury, but the state tried again, and in 2006 a new jury found paint companies liable for the estimated $2.4 billion cost of inspecting and renovating the hundreds of thousands of pre-1978 homes in Rhode Island. Anti-lead advocates around the country hailed the outcome, comparing it with the 1998 settlement in which tobacco companies agreed to pay $206 billion to reimburse states for smoking-related health costs. However, in 2008 the Rhode Island Supreme Court overturned the verdict, finding that public nuisance law isn't applicable in the case of lead paint. The paint companies would not be forced to pay for the cleanup of their product.[27]

Rhode Island wasn't the only place where paint companies were sued over the ongoing lead paint hazard. On the other side of the country, a group of cities and counties in California filed their own lawsuit against major paint manufacturers. A three-judge panel of the state Court of Appeal ruled in 2017 that public nuisance law *is* applicable to the issue of lead paint, and ordered the paint companies to pay approximately $400 million for cleanup. The paint companies fought this result all the way to the U.S. Supreme Court, which declined to overturn the ruling,[28] paving the way for those California cities and counties to collect industry money for lead paint cleanup. Other states are now considering similar lawsuits against the paint industry.[29] These lawsuits represent a critical process for holding the lead industry responsible for the harm it has caused, and demanding that they provide resources for mitigating that harm.

Lead paint and lead pipes are not the only sources that are still exposing kids to lead today—there is also a lot of lead in our soils. As lead researcher Dr. Howard Mielke has pointed out, "We have a Clean Air Act and a Clean Water Act, but no Clean Soil Act."[30] Some of the lead in our soils is a legacy of the many decades of burning leaded gasoline. Since lead is an element, it never breaks down—every atom of lead released from America's tailpipes in the twentieth century is still around somewhere, and most of it settled out in our soils. Soil in large cities, which had the most car and truck traffic during our leaded gasoline years, is especially contaminated.

Dr. Mielke works in New Orleans, where he studies children's exposure to lead from soil. In 2005, when Hurricane Katrina hit New Orleans, the levies failed and many parts of the city were flooded with Mississippi River water, bringing in sediment that settled out and created a new layer of surface soil. The tragedy that cost so many lives and did so much damage also helped to bury some very lead-contaminated soil. Dr. Mielke and his students have shown that surface soil lead concentrations were significantly lower after the storm than before, and these reductions also show up in the blood lead levels of children living in the flooded neighborhoods.[31] His work demonstrates that protecting kids from leaded soil can result in significant reductions in blood lead levels.

All of America's soils are somewhat contaminated from the decades of burning leaded gasoline, but some areas have much more highly contaminated soils from current or past industrial uses. All around the country, there are old industrial sites that used to be lead smelters, or factories making batteries or other lead-containing items, that have contaminated the soil around them. Under federal Superfund law, the companies responsible for this contamination are required to clean it up, but lawsuits to determine financial responsibility can drag on for years, and in some cases the corporation that did the polluting no longer exists, so there's nobody to foot the cleanup bill.[32]

In some cities, old contaminated soils are buried here and there like landmines. Perhaps they have been covered up by new fill or pavement, but as more and more families are moving into urban areas, places that used to be abandoned industrial sites are being redeveloped. These new construction efforts can dig up and disturb highly contaminated soils, releasing dust that exposes children on nearby blocks to high levels of lead. Here in Philadelphia, we once had more lead smelters than any other city in the country. The riverfront neighborhoods where these facilities were located are undergoing gentrification, and the resulting construction is spreading lead-contaminated dust. A 2017 study that tested over one hundred locations in these neighborhoods, including parks and playgrounds, found that 75 percent of them had lead levels considered hazardous to children. In one backyard, the level of lead was more than twenty times the acceptable limit.[33]

In addition to soils, paint, and pipes, lead can be found in a number of unexpected places. The CDC warns of the possibility of lead exposure from improperly inspected toys and toy jewelry coming from other countries. In 2006, a child swallowed a heart-shaped metallic charm from a charm bracelet and died of acute lead poisoning.[34] Imported candy from Mexico has also been found to contain high levels of lead, as do certain folk medicines. "Sindoor" is a red powder that some Hindu and Sikh women wear in the part of their hair; one brand was found to be 87 percent lead.[35] And a study in New Jersey found that artificial turf made with nylon fibers can give off hazardous levels of lead dust as it breaks down over time.[36]

There is also, amazingly, still one type of vehicle burning leaded gasoline in the United States—piston engine airplanes. These are the small propeller planes used by individuals and businesses, typically for short flights. There are currently around 167,000 of these airplanes in the United States, emitting 450 tons of lead into the air every year. Children who live near the 20,000 airports that serve planes using leaded gas have been shown to have higher blood lead levels than children living farther away from these airports.[37] As of June 2019, the Federal Aviation Administration reported that it has "established a rigorous test program to facilitate the evaluation and approval of unleaded fuels that will be environmentally safer than leaded fuels, yet as operationally safe as leaded fuels in the current fleet of piston engine aircraft."[38] Given the monumental harm that leaded fuel has done to American children in the past, one hopes that the FAA can evaluate and approve unleaded fuel for these planes as quickly as possible.

Addressing all of the ongoing sources of childhood lead exposure is critically important. We now know that even relatively low levels of lead can seriously affect the developing brains of young children. We know that being exposed to lead harms their learning, attention, and impulse control, and that these effects are permanent. Implementation of a national universal screening program would test all preschool-age children for lead, so that we can catch cases of lead poisoning as early as possible. However, relying on screening alone uses kids as canaries in the coal mine; truly protecting children from lead would require rigorous testing for potential sources of lead exposure so that they can be found *before* kids get poisoned. Adequate protection would also require fully funding lead abatement programs. Cost-benefit analyses consistently show that a dollar spent getting rid of lead in a child's environment pays off much more than a dollar in future value.[39] Solving these problems is not cheap, but economic analysis demonstrates that it's worth the cost.

Drinking water is already tested for lead, but experts have demonstrated that there are many ways to cheat the system. The state agencies that carry out this testing are supposed to be working under strict oversight from the federal EPA to ensure that they're not fudging their results. As we've seen in Flint, this system can break down without ironclad protections for whistleblowers who speak up when they realize something is amiss. If we're going to address ongoing problems in the drinking water supply, we will need to speed up the decades-long undertaking of replacing the nation's lead pipes. A 2017 report by the Health Impact Project found that "removing leaded drinking water service lines from the homes of children born in 2018 would protect more than 350,000 children and yield $2.7 billion in future benefits, or about $1.33 per dollar invested."[40]

Another critical way to protect kids in the United States from ongoing lead exposure is to find and remove old lead paint. Fully addressing the problem of lead paint would require that every building built before 1978 be inspected, and if lead paint is found, the property would require full abatement, not just cosmetic touch-ups. Full abatement means that all lead paint is entirely removed, or completely secured behind a sturdy new surface. This includes both interior and exterior paint, and not just the paint on the walls; paint on baseboards, door frames, and window frames would need to be replaced as well. The Health Impact Project report calculated that "eradicating lead paint hazards from older homes of children from low-income families would provide $3.5 billion in future benefits, or approximately $1.39 per dollar invested, and protect more than 311,000 children."[41]

Replacing older, lead-painted windows is an excellent example of a win-win situation. Back when lead paint was legal, most windows were the old single-paned style. Both of these things changed in the late 1970s—lead paint was outlawed, and builders began using more efficient double-paned windows. As we've seen, old windows with lead-painted window frames are one of the most serious sources of childhood lead exposure today, because opening and closing those windows produces lead dust that kids end up inhaling or ingesting. Replacing these windows eliminates an important route of childhood lead exposure.

But that's not all. Older, single-paned windows also waste energy. When it's cold outside, a building can lose a lot of heat through these windows, and the furnace has to work harder to keep the building warm. If the building has a gas or oil furnace, that means burning more gas or oil, which produces carbon dioxide that contributes to climate change. Electric heat contributes to climate change as well, because in most of the United States, most electricity comes from burning gas and coal, releasing carbon dioxide into the atmosphere. If every window in this country from before 1978 was replaced, we could protect kids from a lot of lead, and also reduce our national carbon footprint, helping to slow down the warming of the planet. As we work to address climate change and a variety of other important contemporary issues, there are a number of lessons from the rise and fall of leaded gasoline that can be applied to challenges we face today.

8

Lessons from the
Lead Battles

▬ ▬ ▬ ▬ ▬ ▬ ▬ ▬ ▬

Lead was added to gasoline in the 1920s over the objections of public health experts, and for half a century the cars and trucks of the United States pumped more and more lead into the air. A group of brave researchers, activists, and government officials fought for regulations that mandated unleaded gasoline, and the amount of lead that American children were exposed to dropped dramatically. The children who were exposed to more lead had gone on to commit, on average, more violent crimes, and the children exposed to less lead committed less violence. Although the fight against childhood lead exposure is still going on, the fight against leaded automotive fuel is long over, and Americans are healthier and safer because of it. Of course, there are many other areas in which Americans are still struggling to create a healthier and safer future for our children. As we fight these new battles, what can we learn from the history of leaded gasoline?

Combating climate change is one of the most important battles currently under way. Every year, the leaders of countries around the world gather in Davos, Switzerland, for a World Economic Forum meeting, and every year they release a *Global Risks Report*, detailing the threats predicted to have the biggest impact on the world in the next ten years. In 2018, climate change and its consequences topped the list.[1] World leaders are concerned about cyberattacks, nuclear weapons, and terrorism, but most of all they're worried about global climate change.

Understanding the nature of the political debate surrounding carbon emissions and climate change is critical if we're going to move that debate in a productive direction. The story of leaded gasoline has some important lessons to teach us about combating climate change. Just like lead, fossil fuels have been used by humans for thousands of years. Archaeological evidence suggests that people began burning coal more than 3,500 years ago.[2] The ancient Greeks burned coal for metalworking; the Byzantine Empire used flaming oil in warfare; and the ancient Chinese burned natural gas to boil brine for making salt.[3] All these uses of fossil fuels were relatively small-scale until the Industrial Revolution in the 1700s, when coal-powered railroads and steamships began to transport people and materials around the world, and humankind's use of fossil fuels skyrocketed.

As with the use of lead, the burning of fossil fuels has had unintended consequences. Miners and drillers have died in explosions and accidents, or died from diseases contracted through years of ongoing contact with toxic materials. Oil spilled from shipwrecked tankers and leaking pipelines has poisoned lakes, rivers, aquifers, and oceans. Coal-fired power plants have filled the air with toxins like mercury, and with the chemicals that cause acid rain. But the most serious unintended consequence of humanity's fossil fuel binge is global climate change.

Overall, climate change has already warmed the Earth by about a degree and a half Fahrenheit since the 1880s, and this warming is accelerating. By the end of this century, it is predicted that the planet will warm by another 1.5°F to more than 6°F, depending on how quickly we get our carbon emissions under control.[4] In our daily life, those might seem like relatively small numbers; however, on a planetary scale, a few degrees can do a lot of damage. Climate change increases the chances of extreme weather events, which kill thousands every year and destroy critical infrastructure. Rising sea levels threaten coastal cities around the world. Changing temperature and precipitation patterns endanger the global food supply and foster the spread of deadly diseases and the pests that carry them. As temperatures rise, many of the Earth's plant and animal species are threatened with extinction. The effects of climate change are felt from the ice sheets of Antarctica to the malaria wards of tropical hospitals to the produce aisle at your local grocery store.

Some threats to human well-being are highly visible; when there's a car crash, or a terrorist attack, it doesn't take highly trained experts to know that it has happened. Yet both airborne lead and carbon dioxide (CO_2—the main cause of climate change) are invisible. People couldn't see the lead coming out of tailpipes in previous decades, and we can't see the CO_2 being released from tailpipes and smokestacks today, so ordinary citizens are dependent on experts to know that these threats exist. Scientists use highly technical machines to measure the amount of lead or CO_2 released into the air, and scientific research was necessary to determine that lead and other heavy metals harm the brain and that CO_2 and other greenhouse gases warm the planet. Unfortunately, only around 40 percent of Americans have "a great deal of confidence" in the scientific community, according to Pew Research Center polling, a percentage that has remained about the same for the four decades that Pew has been asking the question.[5] Scientists in the 1970s faced a mistrustful public as they attempted to explain the harmful effects of leaded gasoline, and scientists today face a similarly mistrustful public as they work to explain the harmful effects of greenhouse gas emissions.

We won't be successful as a society in undertaking the measures necessary to address climate change if lawmakers and the public are not convinced that human-caused climate change is a real threat. The situation was similar in the fight for unleaded gasoline—scientists struggled to convince politicians and ordinary people that low-level childhood lead exposure was a legitimate concern. In addition to lacking confidence in scientists, many Americans lack basic scientific literacy. A 2015 study rated only 29 percent of Americans as "scientifically literate."[6] Of course, scientific literacy isn't just about knowing the facts; it's also about believing those facts. All of us have a desire to discount information that leads to conclusions we don't like, and that has provided an opening for both the lead industry and the fossil fuel industry to manipulate public perceptions of science.

In chapter 3 we saw a number of examples of how the lead industry attempted to undermine the scientific consensus that low-level lead exposure, including exposure to the by-products of leaded gasoline, was harming the developing brains of young children. For decades, the lead industry carried out its own studies, many of which purported to show the safety of low and moderate levels of lead exposure. When researchers with no ties to the industry started reporting results that conflicted with these claims, the lead industry attempted to discredit the science and the scientists themselves. Dr. Herbert Needleman's research showed that kids who had higher amounts of lead built up in their baby teeth had more learning difficulties and behavior problems in school. A lead industry trade association and scientists with lead industry funding worked hard to find any possible flaws in Dr. Needleman's research.[7] They claimed that there was bias in how he chose his subjects, that there were problems with his statistical analysis, and that there were other factors that he hadn't accounted for.

Dr. Needleman's critics didn't stop at just attacking his studies; they went after the man himself. Industry-affiliated scientists accused him of scientific misconduct on two separate occasions. At one point, misconduct accusations were made to the Environmental Protection Agency to try to convince the agency not to use Dr. Needleman's research in their deliberations about leaded gasoline. These accusations were found to be without merit, and the EPA relied heavily on his work in determining safe lead levels for children. Years later, other scientists on the payroll of the lead industry brought scientific misconduct charges through the University of Pittsburgh, where Dr. Needleman worked. Although he was eventually cleared of any wrongdoing, these accusations and the resulting hearings were very stressful for Dr. Needleman and his family.[8] Also, in the minds of people not intimately familiar with the outcome of such hearings, even the accusation of misconduct can cast a shadow over all of a researcher's work.

The fossil fuel industry has used many of these same strategies to target science and scientists. Climate change deniers continually pick apart scientific articles, arguing that this data point or that statistical technique is in error,[9] trying to use disagreements over trivial details to cast doubt on the entire body of scientific research and undermine the case for immediate and substantial changes to address the threat of global climate change.[10] With hundreds of climate papers being published every year, there's a lot of data out there, so undoubtedly they can find *something* that seems questionable. If you have enough haystacks, you're sure to find some needles. A rule issued by the EPA under the Trump administration facilitates this method of undermining scientific consensus.[11] This rule states that in their decision making on environmental policy, EPA scientists should use only studies whose original raw data sets have been made publicly available.

In addition to providing fodder for disingenuous nitpickers, this EPA rule prevents the agency from using a great many studies whose authors cannot release their raw data to the public, including public health studies in which research subjects were guaranteed that their personal health information would be kept confidential. Many climate studies use data from satellites, and researchers pay for the data; these payments cover the cost of launching and operating the satellites. The researchers are not allowed to make all that expensive data freely available for anybody to use. Imagine if EPA employees were only allowed to read books whose full text was freely available on the Internet—that would exclude a lot of books (including this one), because authors, editors, and publishers need to make a living. EPA officials claimed that this new rule was in the service of "transparency," but scientists see it for what it is—an attempt by the Trump administration to undermine the use of sound science in political policy making, just as some Reagan administration officials worked to undermine sound science about childhood lead exposure.

Another method that climate change deniers have used to undermine valid scientific conclusions also hearkens back to the days of the fight over leaded gasoline—the suggestion of phantom "other" factors. When Dr. Needleman examined the relationship between lead and academic success, he tried to control for as many other influences as he could. However, scientists can never rule out the possibility of an overlooked factor with 100 percent confidence in a single study, and Dr. Needleman's critics kept arguing that there might be something he left out. But study after study, with different methodologies and different subject pools, showed the same conclusion—exposure to lead damages the developing brains of children, even at low doses. Climate change deniers also often point to mysterious "other" factors that might be responsible for the measured rise in global temperatures, raising obscure possibilities like changes in solar energy, or cycles in the Earth's oscillation.[12] No matter how many of these theories are conclusively disproven by climate scientists, there will always be another potential "other factor" waiting in the wings.

Whether it's nitpicking the data or raising phantom yet-to-be-studied influences, these arguments that the science isn't yet certain are often followed by calls for more study. The lead industry argued that more studies were needed to determine the effects of low-level lead exposure,[13] and the fossil fuel industry has consistently argued that more study is needed on the causes and remedies of global climate change. When the EPA issued a proposal to limit carbon emissions from power plants in 2009, a group of fossil fuel companies argued that the science was too uncertain, encouraging the EPA to "withdraw the Proposal, and proceed with caution going forward."[14] In 2017, the Association of Global Automakers demanded that the EPA roll back its newer, more stringent fuel economy standards for vehicles, claiming: "To be clear, we are committed to reducing emissions, improving fuel economy, and bringing carbon-neutral or zero carbon technologies to market. We recognize the need for robust regulation to help ensure consistent progress toward that goal. All we're seeking in this request is a thoughtful, fact-driven exercise that gets us to our goals in the smartest way possible."[15] If the goal is to avoid any regulation of an industry, claiming that we just need some more information first is always a good bet. The world never runs out of information.

You'd think that an industry that's constantly calling for more studies, more science, would be a boon to working scientists. However, just as the lead industry attacked Dr. Needleman and others, the fossil fuel industry and its allies have gone to great lengths not only to challenge the scientific consensus, but also to target climate scientists themselves. The excellent book (and accompanying documentary) *Merchants of Doubt* details how climate-denier think tanks have gone after individual climate scientists, publishing their personal contact information online. Some of these scientists received death threats.[16] There's a reason why climate scientist Dr. Stephen Schneider wrote a book about his experiences entitled *Science as a Contact Sport*.[17] The result is that the average American sees what looks like a legitimate scientific "controversy" rather than a concerted attempt by a few powerful industries to undermine the scientific consensus. Creating doubt is an effective tactic for those promoting inaction—as long as the public believes that the science is uncertain, they will be much less likely to heed calls for immediate, dramatic efforts to address the issue.

The problem is not just that Americans don't know as much as we should about science. The problem is that what we do know about science is strongly influenced by our social, religious, and political beliefs. In the 2015 scientific literacy study mentioned above, respondents did much better on value-neutral questions about the layers of the Earth and the speed of light than they did on more emotionally charged questions about evolution and the origins of the universe.[18] This poses a complex challenge for the scientific community. Just sharing facts with the public is never going to be enough to change their views.[19] If those facts challenge people's deeply held beliefs, they will find a way to discount them, by questioning validity of the science or by questioning the integrity of the scientists themselves.

Back in the 1960s and 1970s, it was becoming clear to many scientists that exposure to lead harms children at much lower levels than we had previously thought, and that significant government intervention was required to protect those children. Oil companies were in the business of making and profiting from leaded gasoline, and they weren't going to stop unless somebody forced them to stop. For those who believe that intervening in business practices to protect the health and well-being of the public is a sacred duty of the government, the situation was clear. The number of studies on low-level lead exposure was still relatively small, but they were all pointing in the same direction. There was enough certainty to act.

However, this was the height of the Cold War, and there were a lot of conservatives in the United States who thought that government intervention in business practices would send us down the slippery slope toward Communism.[20] If the government can tell private companies what products they can and can't manufacture, that takes us one step closer to a Soviet-style central command economy. For people who were politically predisposed to view significant government intervention in private business as a potentially existential threat to American democracy, it would take a great deal more evidence before they would accept a scientific conclusion that pointed toward the necessity of banning leaded gasoline. As the renowned environmental historian Samuel P. Hays would later write about the battles over lead, "Objections to the contemplated policies give rise to challenges to the underlying science. . . . Scientific inquiry as a result becomes controversial at every step because of its continued implications for public policy."[21]

Fears about government overreach and inching toward Communism are still common in our political discourse.[22] While the scientific consensus around low-level lead exposure required some government intervention in the economy, the scientific consensus around the causes and consequences of global climate change suggests the need for intervention at a whole new level. For the anti-government right, this level of federal intervention is unthinkable, so it makes sense that they would be looking for any results that would seem to minimize the amount of necessary government intervention. When different studies came to different conclusions about the potential harms of low-level lead exposure, the lead industry relied heavily on the studies that showed less harm. Today's climate models usually produce a range of possible outcomes, and industry lobbyists often highlight the least-extreme end of this range. They argue, for example, that sea level rise in this century "might be as little as" a few inches, ignoring the fact that the other end of the range of predictions suggests that sea level rise might be as much as several feet.[23]

In addition to minimizing the benefits of government intervention, industry groups have tended to exaggerate the costs of any changes to their business. Oil companies argued that switching over to unleaded gasoline would pose an enormous threat to the U.S. gasoline supply, which was especially worrisome in light of the looming oil crisis. In 1973 the EPA estimated that phasing out leaded gasoline would reduce the country's gasoline supply by about 30,000 barrels per day. The oil industry's estimate was thirty-three times higher—they predicted a reduction of a million barrels per day, and took out a full-page ad in the *New York Times* proclaiming this dire prediction. The ad depicted an oil barrel festooned with an American flag, pouring its contents down the drain.[24]

It's hard to measure exactly how the switch to unleaded affected gasoline supply, because both the Arab oil embargo of 1973 and the Iranian Revolution of 1979 had bigger effects on American oil supplies than anti-lead regulations did. In the end, oil companies managed to switch over to unleaded gas without going out of business, and a study in 1984 suggested that the benefits of the phaseout of leaded gasoline had already exceeded the costs by $700 million.[25] During the second half of the leaded gasoline phaseout, in the late 1980s and early 1990s, oil supplies were so high that many described the situation as a "glut" of oil, and gasoline prices were at historic lows.[26] Clearly, the industry's dire predictions about the impact of switching to unleaded gasoline on the nation's oil supply and gas prices did not come to pass. Years later, the EPA calculated that the benefits of the phaseout had outweighed the costs by ten to one.[27]

The fossil fuel industry has made the same kind of argument against government policies to address global climate change—they argue that reducing carbon emissions would be prohibitively expensive. For example, in 2015 the EPA estimated that the Obama administration's newly enacted Clean Power Plan would result in a 4 percent increase in the cost of producing electricity.[28] The Institute for Energy Research (IER), which is funded by petroleum and coal companies, predicted a whopping 21 percent increase in the cost of producing electricity.[29] However, a 2016 study by the Nicholas Institute for Environmental Policy Solutions at Duke University[30] and a 2017 analysis by the

Institute for Policy Integrity at the New York University School of Law both calculated that costs would be lower than the 4 percent that the EPA had predicted.[31] We will never know the true costs (or benefits) of the Clean Power Plan, because it was dismantled by the Trump administration, but many different researchers exposed the fossil fuel industry's dire prediction of a 21 percent increase in electricity costs for what it was—a scare tactic.

The same scare tactic has been used in predicting the cost of implementing stricter fuel economy standards for America's cars and trucks. In 2016, the EPA projected the cost of its more stringent standards as $1,070 per vehicle.[32] Energy researcher John M. DeCicco at the University of Michigan predicted that the cost would be only $240 per vehicle,[33] and an analysis by *Consumer Reports* found that consumers would actually *save* money because of reduced fuel costs.[34] However, the conservative Heritage Foundation, whose founding board members included automobile and oil company executives, issued a report claiming that "the regulations are adding at least $3,800 (perhaps much more) to the average price of new vehicles."[35] In 2016, as car companies were building vehicles under the new economy standards, employment in the automobile industry in the United States continued to rise, suggesting that the industry wasn't suffering too much.[36] Once again, it seems, a polluting industry had cried wolf, projecting unrealistically high costs for reducing its own pollution. We will never know the final costs or benefits of the more stringent fuel efficiency standards because they have been scrapped by the Trump administration, which called for—guess what?—more research on the issue.

Politicians on the right often argue for getting rid of "job-killing regulations." It is an article of faith in conservative political circles that federal regulations stifle the economy and destroy jobs. For many people, this idea tips the balance in thinking about potential environmental regulations—what good is having a cleaner environment if you can't put food on the table for your family? This is a long-standing concern—way back in the 1920s, corporate executives were arguing that any attempt to regulate leaded gasoline would harm the economy and cost jobs. The problem with this understanding of the relationship between regulation and jobs is that it isn't true.

A major study in 1995 found that environmental regulation had little or no impact on the competitiveness of U.S. manufacturing.[37] In 2011, the *Washington Post* concluded that "economists who have studied the matter say that there is little evidence that regulations cause massive job loss in the economy, and that rolling them back would not lead to a boom in job creation."[38] This conclusion was supported by another study the following year by two economists from the George Washington University.[39] In 2018, the economist Dr. Alex Tabarrok and a colleague published a new study of the effects of federal regulation using a novel machine-learning algorithm.[40] Despite Dr. Tabarrok's libertarian political leanings,[41] his study found that higher rates of federal regulation were associated with slightly *higher* rates of job creation. He and his colleague examined the data every which way, measuring "different subsets of firms, delayed impacts of regulation, different types of regulations and regulatory agencies, measuring the effects of regulation through supply chains, and controlling for measurement error." Still, they could find no evidence that federal regulation kills jobs.[42]

Fossil fuel and automobile companies today, like lead and gasoline companies before them, are wrong when they claim that any new regulations will lead to massive job losses. They are wrong when they claim that complying with safer standards will cost a fortune, and wrong about the costs outweighing the benefits. These industries are also wrong when they try to claim that the harm they are contributing to—brain damage from childhood lead exposure, global climate change from carbon emissions—may not even exist. They have manipulated the science and manipulated the public discourse, but they cannot hide the truth—Americans are safer and healthier now because we banned leaded gasoline, and research consistently shows the benefits of reducing carbon emissions, regardless of what these Chicken Little industries would have us believe.

Climate change regulations aren't the only policies that make our country and our world healthier and safer. The story of leaded gasoline can teach us lessons that apply to many other battles as well. One of those lessons comes from the fact that so few people have heard the good news about the substantial reductions in childhood blood lead levels that followed the banning of leaded gasoline: we need to do a better job of telling our success stories. In my Environmental Science class, I always go around the room on the first day and have students share an environmental issue that they're concerned about. Two decades into the new millennium, some of them are still saying "the ozone layer." But the ozone layer is actually improving, and has been for years. The global treaty that banned the production of CFCs, the primary chemicals responsible for depleting the ozone layer, was signed in 1987. The hole in the ozone layer has been shrinking since before most of these students were born. By the end of the century, the problem should be mostly resolved.[43] My students have no idea.

Environmentalists aren't always good at telling success stories. We live in a country where rivers used to regularly *catch on fire*. It wasn't just the Cuyahoga River in Cleveland—rivers in a number of industrial cities and towns around the country burned in the 1960s.[44] As of 1960, companies could dump anything into rivers as long as it didn't impact "navigability." Similarly, companies could release anything from their smokestacks, and people living downwind were stuck breathing it. Some of those airborne chemicals came back down as acid rain, poisoning forests, lakes, and rivers all over the northeastern United States. In factories and other workplaces, employees had no legal protection from toxic chemicals like asbestos. This isn't ancient history we're talking about; this was within the lifetimes of more than one-third of Americans alive today.[45]

Environmentally, many things have gotten much better. Our rivers have not only stopped catching on fire, but most of them are now safe for fishing and swimming. The Potomac River in Washington, D.C., which had been fouled with raw sewage from antebellum times to the mid-twentieth century, is now a popular spot for kayaking and paddleboarding.[46] Air quality has also improved. According to the EPA, the total emissions of the six main air pollutants that they monitor are now less than one-third of what they used to be.[47] Acid rain levels have dropped in half.[48] Asbestos production has been eliminated in the United States, and many other workplace hazards are tightly regulated[49]—not to mention that the amount of lead in the bodies of American preschool children has declined by more than 90 percent.

Of course, not all environmental problems have been solved, but that doesn't mean we can't celebrate our successes. Success is inspiring. Who wouldn't want to join a movement that has made such incredible progress in cleaning up the environment and protecting human health? Past successes can inspire us as we continue to take on daunting tasks. If the world can come together to solve the problem of ozone depletion, maybe we can work together to make progress in fighting global climate change. Honoring past successes also helps people recognize which problems do, and don't, need our attention in the current moment. If the public were better informed about the problems that have actually been solved, fewer of my students would still be fretting about the hole in the ozone layer.

Perhaps the most important reason for highlighting past environmental successes is to change the debate about government regulations. In a 2016 poll, about a third of Americans said that environmental regulations "cost too many jobs and hurt the economy."[50] Among conservative Republicans, almost two-thirds agreed with that statement. In addition to being more concerned about costs, Republicans are also less confident about the benefits of government regulation. While 76 percent of liberal Democrats believe that restricting power plant emissions can make a big difference in addressing climate change, only 29 percent of conservative Republicans agreed.[51] People generally will not support new government regulations if they don't think they will do any good. However, history shows that environmental regulations have done a world of good, in ways that affect the lives of ordinary Americans every day. Unrestricted free market capitalism didn't give us the safer, healthier environment we now live in; government regulation did.

Clearly, we can't let these environmental improvements make us complacent. As we saw in chapter 7, the problem of childhood lead exposure is far from solved. There is much more work to be done. On the one hand, getting the lead out of gasoline has done an enormous amount of good, including making us all safer from violence than we would otherwise have been. Understanding that success can change the way people think about a lot of things, including the value of environmental regulations. On the other hand, far too many American children are still being exposed to dangerous levels of lead today. The story of the switch to unleaded gasoline doesn't show that the problem of childhood lead exposure is *solved*, but it can help remind us that the problem is *solvable*. By regulating gas, paint, pipes, and so on, we've mostly stopped putting new lead into the environment of young children. Now we need to finish the job of getting rid of all the old lead that's already there.

The (eventually) successful effort to get the lead out of our gasoline highlights the importance of good science. Physical scientists revealed the extent of lead contamination in our environment, and health scientists demonstrated the detrimental effects of low-level lead exposure on developing brains and the impacts of that damage on learning and behavior. Without carefully designed studies, carried out by persistent researchers, Americans would never have known how much of this invisible chemical was in our environment or what it was doing to the nation's children. For too long, lead research had been done almost exclusively by industry-funded scientists who somehow managed to conclude that childhood lead exposure was not a serious problem.[52] Public funding for science is critical for supporting scientific work that furthers the common good.

Today, American public funding for scientific research is in jeopardy. Total federal spending on research and development (R&D), as a percentage of the overall economy, has fallen by around 40 percent in the past three decades.[53] Meanwhile, corporate spending on R&D has skyrocketed. In the 1980s, the amount businesses spent on R&D was roughly equivalent to what the federal government spent. Today, companies are spending more than three times what the U.S. government is. When industries have control over three-quarters of the research and development in the country, the science is likely to reflect corporate interests much more than public interests. There are currently more than 80,000 chemicals in use in the United States, and most of them have not been tested for human health effects.[54] Of the chemicals that have been examined, some have shown cause for concern, but the research is not yet considered conclusive enough to ban them. For example, there are flame retardants and plastic components found in many consumer products that have been linked to cancer and reproductive hazards.[55] Only by putting more resources into funding the independent scientists who are studying these health effects can we understand which chemicals are safe at what levels.

Sound scientific research can also inform laws and regulations on product safety, violence prevention, health care, education, and so on. In nearly every sphere of our lives, science has the potential to help us understand how things work, and how we can make them work better. If we're going to have government policies that are based on a scientific understanding of reality, the last thing we need is policies that directly attack the scientific process. In 1996, thanks to lobbying from the National Rifle Association, Congress passed the Dickey Amendment, which stated that "none of the funds made available for injury prevention and control at the Centers for Disease Control and Prevention (CDC) may be used to advocate or promote gun control."[56] After the passage of this amendment, the federal government basically stopped funding any research into gun violence causes and prevention. Finally, in the wake of the 2018 school shooting in Parkland, Florida, some federal funds were made available to study this important issue.[57] We know that science-based safety policies can work, and not just because of reductions in lead exposure. America's motor vehicle death rate has dropped by more than half in the past half-century as scientists and engineers developed and imple-mented countless improvements to our vehicles and our roadways.[58] Provid-ing the resources to carefully study other issues impacting the safety of Americans can save lives as well.

Research plays an important role in upholding the common good, and cor-porations and their lobbyists know it, but research alone is not enough to make real change. As we saw in the fight against leaded gasoline, changing the world requires activism in a variety of flavors. Neighborhood groups went door-to-door to get kids tested for lead poisoning, and also pushed city health depart-ments to expand their efforts to find and treat lead-poisoned kids. National advocacy organizations undertook campaigns to educate and persuade the pub-lic. For example, the American Lung Association pushed for a reduction in smog through the use of catalytic converters, which helped to bring about the dawn of unleaded gasoline. National advocacy organizations also fought these issues in the courts, as when the Natural Resources Defense Council forced the government to enforce stricter limits on airborne lead.[59] All these different types of activists worked together to push back against the powerful lead and gasoline industries.

For-profit companies are in business to make a profit; it's not Marxist to say that businesses may not choose to act on behalf of the common good without government oversight. As we've seen, many studies suggest that federal regula-tions cause little to no lasting harm to the industries that they're regulating, and the benefits often far outweigh the costs. Yet, in many cases, industries themselves have managed to prevent stricter regulation. In the short term, it can be expensive and inconvenient for companies to change the way they do business, so they have developed highly successful methods for influencing the political process to achieve their anti-regulatory goals. Corporations spend mil-lions on lobbying to influence current policy makers, and they spend even more millions to influence elections and get like-minded policy makers into office in the first place.[60]

Every single issue that requires federal government intervention or oversight to make the world safer and healthier intersects with the issue of campaign finance, and many activists are now coming together to rally for effective campaign finance reform.[61] The history of the rise and fall of leaded gasoline clearly demonstrates that some problems can only be solved through federal government regulations, implemented over the objections of the corporations that created the problem in the first place. The goal of campaign finance reform is to reduce corporate influence on the government and to increase the political will to regulate and monitor corporations to advance the common good.

When it comes to government regulations, the important decision makers are either elected by the voters or appointed by those we elect. The history of leaded gasoline reminds us that elections do matter. In the late 1970s, under President Carter, the phaseout of leaded gasoline was well under way. With the election of President Reagan in 1980, we got an EPA administrator who quickly made it known that she would not enforce the limits on leaded gas.[62] It took a concerted effort by journalists, scientists, and activists around the country, plus the brave actions of some EPA employees, to get the phaseout back on track. It also helped that President Reagan was attempting to "establish a new environmental credibility" in advance of the upcoming 1984 presidential election.[63]

After his election in 2016, President Trump appointed an EPA administrator who was previously best known for suing the EPA itself, and for denying that carbon dioxide is the primary cause of global climate change, claiming erroneously that "there's tremendous disagreement" on the topic.[64] President Trump later appointed a new administrator who had previously worked as a coal industry lobbyist. During the Trump administration, EPA administrators worked to undo regulations that protect the health and safety of Americans, and to undermine the use of sound science by employees within the agency. Trump's opponent in 2016, Hillary Clinton, called for stricter environmental regulations and increased scientific research. A postelection analysis by the *Washington Post* found that the 2016 election was essentially decided by around 100,000 people in three states—Pennsylvania, Michigan, and Wisconsin.[65] That's a small enough number of people to fit in a single stadium to watch the Michigan Wolverines play football. Every vote really does matter. The history of leaded gasoline has shown us that the person sitting in the Oval Office can have a big impact on which rules get enforced, and whether our government is working to protect the health of its citizens or the bottom line of corporations. For those of us who care about leaving a safer, healthier world for our children, there may be no more important place to focus our energy than the voting booth.

Elections also have an important impact on criminal justice policy. The rise and fall of leaded gasoline and the later rise and fall in the violent crime rate hold some important lessons for how we think about our criminal justice system. Back in the 1990s, when a lead-poisoned generation of adolescents and young adults was committing a record number of violent crimes, our country implemented a number of draconian criminal justice policies. That era saw the rise in popularity of the "broken windows" theory of criminology, which holds that visible signs of criminal activity—such as graffiti or public drunkenness—encourage further criminal activity, including serious crimes.[66] In response, police departments in many cities implemented "zero tolerance" policies toward minor crimes like vandalism and jumping the subway turnstiles. In theory, these policies were designed to deter major crime by reducing the prevalence of lesser crimes.

In practice, this kind of zero tolerance policing has led to high levels of police harassment, and to the arrest and prosecution of millions of Americans for low-level, nonviolent offenses. For example, the American Civil Liberties Union estimates that individuals in New York City have been stopped and questioned by police officers more than five million times since 2002.[67] Black and Latino New Yorkers are especially likely to be targeted by the city's stop-and-frisk policies, aspects of which have been found unconstitutional. These interactions with the police can irrevocably change a person's life. For example, an arrest for drug possession can affect federal student aid, food stamps, and housing assistance. Our cities' decision to "get tough" on low-level crime has had a catastrophic effect on many, many lives.

It wasn't just policing that got harsher during the high-crime era; it was also sentencing. During the decades of rising crime rates, mandatory minimum sentencing was applied to a number of drug offenses. Then, in 1994, the U.S. Department of Justice implemented a "three strikes" provision, some version of which has also been enacted in twenty-six states.[68] These laws were intended to impose a mandatory life sentence on anybody convicted of three separate serious crimes, though the details of which crimes count as a "strike" can be tricky. Americans have been sentenced to life in prison for stealing socks, a slice of pizza, or baby shoes.[69]

The result of all of this aggressive policing and harsh sentencing has been an unprecedented rise in the number of incarcerated Americans. In the mid-1970s, less than half a million Americans were in prison or jail. By the mid-2000s, that number was nearly two and a half million.[70] The United States contains 4.4 percent of the world's population and 22 percent of its prisoners.[71] As with aggressive policing, harsh sentencing disproportionately affects Black Americans. One study found that, on average, Black men serve sentences that are 19 percent longer than the sentences served by White men for the same crime.[72]

Now that crime rates have fallen so dramatically—thanks to numerous factors, including the switch to unleaded gas—the policing and sentencing policies put into place in the high-crime 1990s need to be re-examined. The fight against mass incarceration is one of those rare political issues that can bring together people from all parts of the political spectrum. Not only does mass incarceration tear apart families and ravage communities; it's also incredibly expensive. U.S. taxpayers currently spend about $80 billion every year to keep people in our jails and prisons, a figure that enrages social liberals and fiscal conservatives alike.[73] There's a broad coalition working in states and the federal government to enact sentencing reform.

For me personally, studying the influence of childhood lead exposure on crime rates has highlighted the fact that, in addition to policy changes, our country needs a change of heart. Millions of children's brains were damaged by a factor beyond their control, in ways that have made it more difficult for them to learn, pay attention, and control their impulses. Anybody who ever drove or rode in a car burning leaded gasoline bears some responsibility for this damage. Of course, it's not just lead—many other external factors (chemical, physical, and social) can affect brain development in detrimental ways. Hopefully, recognizing these challenges can help us to have more compassion. People need to be held responsible for their actions, but there are numerous ways to reduce violence and other criminal activity without resorting to a "lock 'em up" mentality. Perhaps acknowledging the impacts of lead and other external factors can help us learn to see a teenager who commits a crime as somebody who needs support and redirection, rather than as a monster.

Many Americans are unaware that we're living in a much safer country now than we were back in the 1990s, and reducing the fear of crime could have a number of advantages. When we're not afraid, we get out and about more, and get to know our neighbors; these positive social interactions have proven health advantages. Creating a feeling of safety contributes to a positive feedback loop in which more "eyes on the street" actually help to make our neighborhoods even safer. People who feel safe walk and use public transportation more, reducing their carbon footprints. When the fear of crime is reduced, parents give their children more freedom, helping them grow up to be less anxious, more independent adults. Sadly, our country still has some very dangerous neighborhoods, but most Americans live in places that are safer than they have been for decades. Perhaps if more of us knew this, more of us would go out and play.

In the end, the story of the rise and fall of leaded gasoline is a story of amazing progress. The scientists and activists fighting against leaded gasoline knew that the free market alone could not solve the problem, so they went up against a powerful and well-funded industry to force the federal government to act. It took decades of hard work. The results, though far from complete, have been amazing. We got the lead out of our gasoline, and more than 90 percent of the lead out of our preschool children. Today's children and young adults are healthier, smarter, and less violent than they would have been if not for that extraordinary effort. It was a long fight for the people involved, and there must have been times when it looked like they would not prevail. Industry lies were taken seriously; anti-lead researchers came under attack; Americans elected an aggressively anti-regulatory president. The switch to unleaded gas was never a foregone conclusion, but the country got there in the end.

In the past few years, many activists have watched past successes being renegotiated, and watched years of progress stripped away. Things can seem bleak. For me, researching the story of leaded gasoline has helped me to focus on the larger sweep of history. Many of us have heard Dr. Martin Luther King Jr.'s much-quoted assertion that "the arc of the moral universe is long, but it bends toward justice." Dr. King was paraphrasing a sermon by the nineteenth-century Unitarian minister Theodore Parker.

Reverend Parker was a social justice warrior way back in the 1840s and 1850s. According to his biographer, John White Chadwick, Reverend Parker was involved in a number of reform movements, including "peace, temperance, education, the condition of women, penal legislation, prison discipline, the moral and mental destitution of the rich, [and] the physical destitution of the poor."[74] Most of all, he advocated for the abolition of slavery. Reverend Parker sheltered runaway slaves in his home, violating the Fugitive Slave Law of 1850. He supported the abolitionist revolutionary John Brown, who was executed. Theodore Parker was a man who lived through challenging times for a cause he believed in. Yet he did not give up hope. Dr. King's famous quote comes from Reverend Parker's sermon "Justice and the Conscience," which he preached in 1852, in the midst of those dark antebellum years: "Look at the facts of the world. You see a continual and progressive triumph of the right. I do not pretend to understand the moral universe; the arc is a long one, my eye reaches but little ways; I cannot calculate the curve and complete the figure by the experience of sight; I can divine it by conscience. And from what I see I am sure it bends towards justice."[75]

It's easy for us to get caught up in the day-to-day struggle, keeping our heads down and our noses to the grindstone. But it's important to look around once in a while and notice how far we've come. Learning about the story of leaded gasoline has convinced me that we must never give up hope. If the arc of the moral universe does bend toward justice, it doesn't bend on its own. It bends toward justice because so many people in the generations that came before us spent their entire lives pushing it in that direction. The arc of the moral universe bends toward justice because so many people get up every day and push toward justice. We can carry the lessons of the past with us as we march toward a brighter future.

Conclusion

— — — — — — — — —

Understanding Our
Leaded World

Our nation's decisions, first to burn leaded gasoline for half a century, and then to switch to unleaded gas, influenced the neurological development of American children for over half a century. Exposure to lead meant that kids born in the 1950s, 1960s, and 1970s were at higher risk for cognitive and behavioral problems than their parents were, and committed significantly more violent crimes as teenagers and young adults. Kids born since 1980 have been exposed to less lead, on average, than previous generations were, reducing their risk of developing these cognitive and behavioral issues, and contributing to significantly lower rates of violent crime.

I wrote this book because I think that those three sentences represent, in a nutshell, an important story that ought to be more well known. In the previous chapter, we saw how the lessons learned from this story can help to inform the efforts of activists working across a range of issues to make the world safer and healthier for future generations. Activists aren't the only ones with a lot to learn from the dramatic story of leaded gasoline—this story has important ramifications for a multitude of topics across the natural sciences, social sciences, and humanities. Scholars in numerous fields may find that this story raises interesting questions and may provide valuable insights that can inform their work.

In my own academic field of environmental science, the story of leaded gasoline highlights the importance of the interdisciplinary nature of our discipline. The true nature of the problem would not have been understood without the work of Clair Patterson, a geochemist who set out to measure the age of the Earth and incidentally discovered just how lead-poisoned our environment was. The impacts of the problem would not have been uncovered without biologists who demonstrated how lead harms our bodies and brains, and physician researchers who showed how that harm leads to long-term, real-world consequences.

Actually addressing the issue of leaded gasoline required environmental scientists to step outside their role as scientists and into the (sometimes uncomfortable) realm of advocacy. The story of leaded gasoline is an important case study for examining a number of questions at the intersection between environmental science and public policy. How much certainty is required before we take action to address a potential environmental hazard? Whose scientific conclusions are credible, and whose are biased by their affiliations or funding sources? How do we make sure vulnerable populations are protected from toxic chemicals? All of these questions are also being asked about the environmental threats we face today, and understanding the history of lead can influence our thinking about current challenges in environmental science, including the ongoing challenge of childhood lead exposure here in the United States and around the world.

The story of leaded gasoline also highlights the importance of advocacy by public health professionals. Public health experts in this story were like the character at the beginning of the horror movie who says, "I don't think we should go into that spooky old house." When leaded gas was first introduced in 1924, a number of prominent public health experts warned of its potential dangers.[1] Public health professionals were the heroes of the next part of the story as well. As early as the 1950s, there was some excellent work being done in Baltimore finding and treating cases of childhood lead poisoning. By the 1960s, public health departments in a number of major cities were tackling the problem of childhood lead exposure, and this work at the local level—neighborhood by neighborhood, block by block—helped to propel the issue to national attention.

The story of leaded gasoline highlights the importance of public health work at a wide variety of scales, from the neighborhood level all the way up to national and international advocacy. (Many countries outside of North America and Europe have been quite slow to get the lead out of their gasoline.[2]) It also demonstrates that the effects of a successful public health campaign can go far beyond what we typically think of as "health" outcomes. Because toxins in the environment can affect brains and behavior, kids growing up in less toxic homes and neighborhoods are likely not only to be healthier, but also to have better educational, social, and employment outcomes as well. Pointing to those secondary effects can be a helpful way to advocate for increased public health resources.

The story of leaded gasoline highlights the importance of brain development in utero and during the first few years of life, and how much damage can be done to developing brains by the presence of a toxic chemical. Neuroscientists now understand a great deal about these critical early developmental processes and the ways in which they can be disrupted. We now know that a brain that spends its first several years soaking in a neurotoxic environment can suffer permanent damage, the kind that alters cognition and emotional processing throughout that person's life.

The issue of lead and crime speaks to a number of crucial questions about brain and behavior. We know that certain structural and biochemical aspects of the brain are *correlated* with certain behavioral outcomes, but to what degree is brain chemistry destiny? Neuroscientists have demonstrated that childhood lead exposure damages the brain, causing changes that affect learning, attention, and impulse control. This story emphasizes the challenges involved in applying neuroscience information to our understanding of history, of society, of crime and punishment. Lead exposure is just one of the many ways that our environment impacts the development of our brains, and neuroscientists keep discovering new environmental influences on brain function all the time. As a society, we must grapple with what those discoveries mean for how we think, how we behave, and who we are.

Of course, we could not understand the story of the rise and fall of leaded gasoline without a historical understanding of who did what, and why, at different points in history. This story is steeped in the attitudes, beliefs, and practices during different eras of the twentieth century, and touches so many aspects of our history, from changing crime rates to why barns are tradition- ally red. (White lead pigment was relatively expensive, so people saved it for the farmhouse, but barns still had to be painted to protect against rot, so people used cheaper red iron pigment instead.[3]) Every historical narrative connects to so many others. The design of early automobiles connects to gender roles—the crank starters on early automobiles were so difficult to turn that typically only men could do it, so the invention of the electric starter struck a blow for women's liberation.[4] Racial politics and public health intersect in the story of a group of Young Lords (the Puerto Rican counterpart to the Black Panthers) going door-to-door in New York City neighborhoods in 1969, collecting urine samples from young children to test for lead poisoning.[5] Like so many other historical stories, this one high- lights the importance of bringing together political, economic, social, and scientific perspectives.

The history of leaded gasoline is also an excellent illustration of the impact of the past on the present. For example, the fight against mass incarceration is often in the headlines these days. Mass incarceration exists in this country to a large degree because of changes that were made to our criminal justice system in the last two decades of the twentieth century. Those changes were made as a fearful voting public reacted to skyrocketing crime rates. Crime rates had been rising for decades as a result of a number of social and economic factors, which were exacerbated by an ongoing rise in childhood lead exposure that can be traced back to the introduction of leaded gasoline in 1924. Our politics, laws, and culture are always shaped by historical events, and this story highlights many of those historical connections.

The story of leaded gasoline also provides valuable insights in the field of economics, especially about the use of cost-benefit analysis. It was understood as far back as the 1920s that there were some health costs associated with the use of leaded gasoline. A number of workers had died in the manufacturing plants that made leaded gas, so at the very least, companies had to consider the cost of replacing those workers. However, at that time, the automobile and oil industries were able to convince government regulators and the public that the benefits far outweighed the costs. Leaded gas would allow cars to be bigger, more powerful, and more fuel-efficient. In a country that was just starting to take to the open road in earnest, those were compelling benefits to consider. So the Ethyl Corporation made billions selling leaded gas, and the widespread public health effects were left as externalities, costs to be paid by the rest of us for decades to come.

The fight in the 1960s and 1970s to remove the lead from our gas was also a fight about cost-benefit analysis. Industry experts claimed that the health costs were too small, or too uncertain, to outweigh the ongoing, demonstrated benefits of leaded gasoline for our economy. Eventually, the anti-lead side prevailed, partially by developing a mountain of evidence to back up their claims that the health costs of leaded gasoline were far greater than industry analyses suggested. The lead industry had also (surprise, surprise) overestimated the cost of switching to unleaded gas. The cost-benefit analysis that was carried out by EPA economist Joel Schwartz, often considered the first major cost-benefit analysis of a health policy, suggested that the benefits of a rapid phaseout of leaded gasoline would far outweigh the costs.

An economist colleague once mentioned to me that economists spend a lot of time carrying out analyses of future costs and benefits, but not a lot of time looking back at past cost-benefit analyses to see how they turned out. In 1984, Joel Schwartz calculated that the benefits of the phaseout of leaded gas would outweigh the costs by about two to one.[6] His work convinced regulators, and by 1996 leaded gasoline had been phased out completely. In 1997 a new EPA analysis calculated that the benefits had actually outweighed the costs by ten to one.[7] Although the original analysis had underestimated some of the costs of phasing out leaded gasoline, it had also greatly underestimated the benefits. The leaded gasoline phaseout provides an excellent case study in both the advantages and challenges of cost-benefit analysis.

The story of leaded gasoline also provides an excellent example of an economic externality, a situation in which part of the cost of making or using a product (leaded gas) is paid for by somebody other than the producer of the product (everybody who was breathing the air). The only reason that selling leaded gasoline in the 1950s and 1960s was cheaper than switching to selling unleaded was because the health, educational, and employment costs of poisoning a generation of children weren't being paid by the oil companies themselves. Forcing companies to stop creating those externalized costs required federal government intervention, and the outcomes of that intervention turned out to be overwhelmingly positive. As we work to address complex economic challenges in the future, including the externalities associated with fossil fuel use, it is valuable to look back at how one important economic issue played out in the past.

The history of leaded gasoline also raises a number of political science issues, including the impacts of the changing role of the government in our society. Back in 1924, when leaded gas was first introduced, most government agencies at the local, state, and federal levels were quite friendly to the lead industry. Many people thought that so as long as companies were competing fairly with each other, there was no cause for government intervention in the economy. In fact, it's not at all clear that the federal government had the regulatory authority to ban leaded gasoline even if they'd wanted to. If workers in the factories that were producing leaded gasoline sometimes suffered from severe lead poisoning, well, that was between the employees and their employers. Government agencies could make recommendations, but had very little enforcement power.

As political scientists know, the scope and power of the federal government had changed by the 1970s. The rise of the environmental movement had highlighted a role for the federal government in protecting the health of its citizens and their surroundings. These policies didn't only protect the environment—government also became much more active in protecting workers' rights, civil rights, and product safety. In the single decade from 1968 to 1978, Congress passed more regulatory statutes than it had in the country's entire previous 179-year history.[8] This newly expanded role of government allowed for regulations that banned leaded gasoline. The expanding power of the federal judiciary played a role as well. When the regulatory process stalled, advocacy groups took their case to court, and sympathetic judges ordered the implementation of new regulations without delay.

The story of leaded gasoline is also a story of whose voices get heard in the halls of government. For a long time, industry scientists dominated the discussion about any possible health hazards associated with leaded gasoline. When other scientists came along with a different story to tell, the lead industry worked hard to discredit them, sometimes attacking them personally and professionally. Nevertheless, they persisted, and over time a coalition of scientists, activists, and the public was able to get through to government decision makers and get the laws changed. This story is relevant to contemporary issues of free speech, influence peddling, and democracy.

The significant evidence that childhood lead exposure can affect later crime rates means that the story of leaded gasoline also has important ramifications for the field of criminology. Criminologists study numerous factors that influence crime rates—social, psychological, familial, cultural, economic, and so on. Unfortunately, the study of biological influences on criminal behavior has a sordid past. The phrenologists and eugenicists of yore still cast a long shadow over today's investigations of brains and crime, and criminologists know enough to be on guard for any argument that criminals "can't help it" because of their biology. However, some criminologists reject childhood lead exposure as a relevant factor simply because their research is focused elsewhere.

For example, in his 2018 book *Uneasy Peace: The Great Crime Decline, the Renewal of City Life, and the Next War on Violence*, criminologist Patrick Sharkey argues that it was "the hard work of community groups, combined with the enhanced presence of law enforcement, the criminal justice system, and private security forces," that caused violent crime rates to fall after the mid-1990s.[9] In the book's one and only paragraph on lead, Sharkey concedes that there is "compelling evidence" that childhood lead exposure can influence the chances that a person will later commit a crime, but he concludes that declining lead exposure did not play a "major role" in declining crime rates. He goes on to dismiss the importance of all other factors that were not direct responses by American society to the problem of violent crime.[10] Sharkey's focus is on the ways that Americans responded to the crime epidemic, and so any factors that do not fit into that category are given the brush-off. A 2015 review by the Brennan Center for Justice similarly recognizes the existence of a relationship between childhood lead exposure and violent crime rates, but dismisses the importance of reduced lead exposure in reducing crime.[11]

Back in 1997, criminologists Franklin Zimring and Gordon Hawkins warned their colleagues against a tendency to focus on only one or a few causes of violent crime, and to "dismiss as insignificant any factor other than that which the speaker is nominating."[12] Zimring and Hawkins argue that "a large number of necessary causes can be viewed as coexistent at different levels of explanation and in no logical sense can they be regarded as competitive with one another."[13] Highlighting the importance of childhood lead exposure does not diminish the importance of other factors; it simply adds an additional factor into consideration. Arguably, leaded gasoline was not just a cause of violence; it was itself a form of systemic violence done to America's children, and not all children were affected equally. The influences of poverty, trauma, and social disruption have been exacerbated by the prevalence of childhood lead exposure.

This story adds another shade to the already vastly complex set of interacting factors that influence who does and does not engage in violent and/or "criminal" behavior.

In addition to issues of criminal behavior, the story of leaded gasoline also raises a number of other sociological questions. Why were so many people primed to believe the industry experts who claimed that leaded gas wouldn't hurt us? Our understanding of who is or is not a credible "expert" has many ramifications for today's policy debates. What role did racism and classism play in the dismissal of lead poisoning during the decades when it was considered a "ghetto problem"? Sexism fits in here as well, since some doctors argued that little kids were only eating paint chips because their selfish working mothers weren't supervising them closely enough. Why are so many of us resistant to the notion that an environmental toxin like lead could alter our behavior? We can accept the idea that invisible chemicals in our environment can affect our *physical* health, but it's harder to imagine them influencing our thoughts and feelings.

One particular sociological question that has broad implications for American life today is this: Why have so few people noticed that violent crime rates have dropped in half over the past few decades? Year after year, I teach classes of university students who are convinced that they're living in a uniquely dangerous time. But the facts are clear—an American living in the 2020s is far less likely to be the victim of a violent crime than an American living in the 1990s was. My students are living in a media-saturated age, and nothing commands attention like the breathless coverage of violence and all the warnings that go along with that coverage. ("Could a predator be living in your neighborhood? Stay tuned!") When that constant background hum of fear takes over, it can be hard for facts to work their way into our brains.

Perhaps another reason it is so difficult to convince people that violent crime rates are much lower now than they were thirty years ago (and to alter their behavior and life choices accordingly) is a fear that crime rates could go back up at any time. Of course, with so many factors influencing rates of violent crime, it's difficult to forecast future crime rates. However, the evidence suggests that leaded gasoline was one of the factors that contributed to the crime epidemic of the 1980s and 1990s, and leaded gasoline is gone now. Could knowledge about the relationship between childhood lead exposure and violent crime help people feel more confident about the safer world we live in today?

The story of changing levels of childhood lead exposure and changing rates of criminal behavior also touches on some important philosophical issues. Philosophers ask big questions, and one of the biggest is about whether and to what degree we, as individuals, have free will. Along with questions about free will come questions about blameworthiness and punishment, and this story raises some of these questions explicitly. If a neurotoxin in your environment can alter your brain in such a way as to nudge you toward violent and/or "criminal" behavior, did you really choose that behavior? Can you be held accountable for it? Does anybody else (oil companies? car and truck drivers?) share in the responsibility for your behavior? If we, as a society, have failed in our responsibility to protect children from neurological harm, do we all share some responsibility for their later behavior? What does that mean for our ideas about crime, punishment, and rehabilitation? This story provides a good opportunity for examining some of these questions.

The story of rising and falling childhood lead exposure raises a number of other philosophical questions as well. The lead industry argued for years that no limitations should be put on leaded gasoline unless scientists could prove definitively that it was harming children. This brings up questions about the nature of scientific inquiry and uncertainty, and how to use imperfect information in decision making. In addition, Americans pumped leaded gas into their cars for decades without realizing that they were doing irreparable harm to our nation's children. This raises questions about the relationship between information and ethics—do we have to know we are doing harm to be held responsible for it? What is our level of obligation to seek out that knowledge? What are we doing right now that may be harming others without realizing it?

There are many ways that knowing the story of leaded gasoline can help us understand our world more clearly. When people around us are debating various policing tactics and criminal justice policies, we know that many of those policies started at a time when violent crime was much more of a threat than it is now, especially in inner cities. When a political candidate promises to slash government regulations, we know how some of those regulations have made us all safer and healthier. When industries claim that changing their practices to protect health and the environment would be impossible or prohibitively expensive, we know that such arguments in the past have often turned out to be overblown.

Understanding the serious and lasting consequences of childhood lead exposure should also inspire us to continue the fight against lead, both here in the United States and abroad. Right now, over half a million American kids have too much lead in their bodies, many of them Black kids, whose lives have too often been undervalued by the political establishment. We can address this ongoing case of environmental racism. Studies have shown that money spent on reducing the amount of lead that kids are exposed to has an excellent return on investment—we just need to make lead abatement a priority. In many developing countries, the level of childhood lead exposure is even higher, and we have an obligation to those kids as well. The United States has a history of shipping health hazards (cigarettes, toxic pesticides, etc.) overseas, and leaded gasoline is just one more example. We should help to clean up the messes we've helped to make.

The story of the rise and fall of leaded gasoline in the United States goes far beyond gas pumps and gas tanks. It is a story that has affected people of different generations and different races all across the United States. It is a story that can influence how we think about a wide variety of subjects in the natural sciences, social sciences, and humanities, and in our daily lives. It is also a story that sheds light on many important contemporary issues, and how we can continue the work to make the world safer and healthier for ourselves and for future generations.

Acknowledgments

As a reader, I always find the acknowledgments section to be an affirming reminder of the interdependent nature of human existence. There may only be one name on the front cover of a book, but there are always lots more names waiting in the acknowledgments section to remind us that nobody truly does anything alone. The opportunity to thank some of the many people who have helped, supported, and encouraged me over the more than five years of this project—and to remind readers that each of us is part of a vast, interdependent web—fills my heart with joy.

When I needed gentle, constructive feedback on the first draft of this book, I reached out to my amazing community at Main Line Unitarian Church. Big thanks to Mary Stromquist, Craig Farr, Rich Fritzson, Kary Heller, Kent Zehner, Lee Dastur, and especially Terry Dixon for the feedback and for believing in me.

My kind and generous colleagues at Cabrini University could not have been more supportive. Thanks to the 2017–2018 Cabrini 1976 Foundation Faculty Fellows for your helpful insights and gentle redirection: Lisa Ratmansky, Nancy Watterson, Amber Gentile, Angela Campbell, Michelle Szpara, Saleem Brown, and Susan Pierson. I want to thank the brilliant criminologists Katie Farina and Bethany Van Brown for being willing to sit down and talk about criminology with somebody who didn't know anything about their field. Please don't hold any of my remaining ignorance against them.

My dear friend Anne Coleman was my department chair when I started this project, and without her support and encouragement the whole thing never would have gotten off the ground. Thanks also to Anne's talented graphic designer son, Shawntrell Coleman, for the image of a neuron in chapter 5. Jeff Gingerich went above and beyond as provost when I told him I would need to replace my sabbatical with maternity leave, and he simply moved my sabbatical to the following year without hesitation. My colleague Eric Malm had recently trod the winding path of academic book publishing, and acted as a guide and mentor.

Nobody could do any scholarly work without the support of a dedicated library staff. Many thanks to library director Anne Schwelm, whose advice about publishing and publishers was indispensable, and to Adam Altman, my interlibrary loan guru. I also want to thank all of my 2019 and 2020 Environmental Justice students. Learning alongside you was a joy, and I especially appreciate those of you who provided detailed and thoughtful responses to the anonymous feedback survey—it was clear that you weren't just doing it for the extra credit points.

Penn professor Rich Pepino was kind enough to share some of his experiences from decades in the environmental justice trenches, and a perspective on childhood lead exposure that comes from many years of fighting the good fight. I also want to thank environmental health advocate Patrick MacRoy, drinking water expert Adrienne Katner, and historian Amanda Seligman for their thoughts, suggestions, and support.

Howard Mielke is a giant in the history of fighting childhood lead poisoning, and he was extremely generous with his time, first when I followed him around at an EPA conference in Philadelphia that I wasn't really invited to, and later when he took a whole day to show me around New Orleans and chat about lead, science, history, and policy. Marc Edwards had recently become world-famous for his work in Flint, Michigan, when I asked to interview him, and he still found time to sit down with me and fill me in on some harsh truths about the state of environmental regulation in America. I am grateful to economist Rick Nevin, not only for letting me use his striking graph in chapter 6, but also for two decades of raising the alarm about the impacts of childhood lead exposure on brains, behavior, and crime.

I want to thank my editor at Rutgers University Press, Peter Mickulas, for taking a chance on a manuscript that was too informal to be a typical academic book but too scholarly to be a typical popular book. I also owe big thanks to my production editor, Sherry Gerstein, who was incredibly patient with the many questions of a first-time book author, and my copyeditor, Barbara Goodhouse, who knows the Chicago Manual of Style like nobody I have ever met. I cannot express how grateful I am for David Rosner—the fact that such a prominent public health historian would take the time to carry on a substantial correspondence, and even sit down with me to discuss revisions, is downright heartwarming. Thanks also to the brilliant and esteemed Gerald Markowitz for his thoughtful feedback and encouragement. I want to thank all the anonymous readers who reviewed my manuscript; I may not have enjoyed receiving their critique, but every bit of it made the book stronger. Thanks, also, to the many editors who sent me lovely, encouraging rejection letters—I really appreciate them taking the time to say something more than "no."

My parents, Jann and Pat Nielsen, could not have been more supportive and encouraging throughout this whole process. They believed in me when I wasn't sure of myself, they took care of my daughters when I needed time to write, and they drank champagne with me to celebrate every tiny step forward on this long journey. I also want to express endless gratitude and love to my incredible daughters, Celia and Anya, who inspire me every day with their curiosity, bravery, and heart.

None of this would have been remotely possible without my partner and the love of my life, Bill D'Agostino. Even in the midst of our busy days, with jobs and kids and a household to run, Bill was always willing to share his remarkable intellect, his keen eye for narrative clarity, and his unwavering faith in me. I could not have gotten through this long process without his comforting hugs, his compassionate co-parenting of our children, and his resolute willingness to put up with me. There is nobody I'd rather be on this adventure with, and I look forward to all our adventures to come. No matter what.

Finally, I want to express my gratitude to you, my reader. Thanks for hanging in there.

Notes

Preface

1 McQuaid, "Without These Whistleblowers, We May Never Have Known the Full Extent of the Flint Water Crisis."
2 Quoted in Lynch, "Whistle-Blower Del Toral Grew Tired of EPA 'Cesspool.'"
3 McQuaid, "Without These Whistleblowers, We May Never Have Known the Full Extent of the Flint Water Crisis."
4 Fortin, "Michigan Will No Longer Provide Free Bottled Water to Flint."
5 Hanna-Attisha et al., "Elevated Blood Lead Levels in Children Associated with the Flint Drinking Water Crisis."
6 Annest and Mahaffey, *Blood Lead Levels for Persons Ages 6 Months–74 Years.*
7 Quoted in Lynch, "Whistle-Blower Del Toral Grew Tired of EPA 'Cesspool.'"
8 Lurie, "Meet the Mom Who Helped Expose Flint's Toxic Water Nightmare."
9 Marc Edwards (Virginia Tech Engineering Professor) in discussion with the author, October 17, 2016.
10 Denworth, *Toxic Truth.*
11 National Minerals Information Center, United States Geological Survey, "Historical Statistics for Mineral Commodities in the United States, Data Series 2005-140."
12 Centers for Disease Control and Prevention, "Blood Lead Levels in Children Aged 1–5 Years—United States, 1999–2010."
13 See chapters 5 and 6 for more details.

Chapter 1   Lead in Twentieth-Century America

1 Mielke, "Dynamic Geochemistry of Tetraethyl Lead Dust during the 20th Century"; Rosner and Markowitz, "A 'Gift of God'?"
2 See chapter 2 for more details.
3 See chapter 3 for more details.
4 Nevin, "How Lead Exposure Relates to Temporal Changes in IQ, Violent Crime, and Unwed Pregnancy."
5 Nevin, *Lucifer Curves.*

6   Benfer, "Contaminated Childhood."
7   Rothstein, *The Color of Law*.
8   See chapter 4 for more details.
9   Warren, *Brush with Death*.
10 National Minerals Information Center, United States Geological Survey, "Historical Statistics for Mineral Commodities in the United States, Data Series 2005-140."
11 National Minerals Information Center, United States Geological Survey, "Historical Statistics for Mineral Commodities in the United States, Data Series 2005–140."
12 Annest and Mahaffey, *Blood Lead Levels for Persons Ages 6 Months–74 Years*.
13 Mattuck et al., "Recent Trends in Childhood Blood Lead Levels."
14 Van Ulirsch et al., "Evaluating and Regulating Lead in Synthetic Turf."
15 Centers for Disease Control and Prevention, "Blood Lead Levels in Children Aged 1–5 Years—United States, 1999–2010."
16 National Minerals Information Center, United States Geological Survey, "Historical Statistics for Mineral Commodities in the United States, Data Series 2005–140."
17 Hillier, "Redlining and the Home Owners' Loan Corporation."
18 Rothstein, *The Color of Law*.
19 Ewens, Tomlin, and Choon Wang, "Statistical Discrimination or Prejudice?"; Bayer et al., "Racial and Ethnic Price Differentials in the Housing Market."
20 Connerly, "From Racial Zoning to Community Empowerment."
21 Annest and Mahaffey, *Blood Lead Levels for Persons Ages 6 Months–74 Years*; Centers for Disease Control and Prevention, "Blood Lead Levels in Children Aged 1–5 Years—United States, 1999–2010."
22 Jee-Lyn García and Sharif, "Black Lives Matter."
23 Centers for Disease Control and Prevention, "Blood Lead Levels in Children Aged 1–5 Years—United States, 1999–2010."
24 Laidlaw et al., "Children's Blood Lead Seasonality in Flint, Michigan."
25 Feigenbaum and Muller, "Lead Exposure and Violent Crime in the Early Twentieth Century."
26 Mielke, "Dynamic Geochemistry of Tetraethyl Lead Dust during the 20th Century."

Chapter 2   Where the Lead Came From

1   Sohn, "Lead."
2   Delile et al., "Lead in Ancient Rome's City Waters."
3   Retief and Cilliers, "Lead Poisoning in Ancient Rome."
4   Sumner, "ScienceShot"; Retief and Cilliers, "Lead Poisoning in Ancient Rome."
5   John of Ephesus, "Ecclesiastical History, Part 3—Book 3."
6   N. Goodman, *The Ingenious Dr. Franklin*, 29–32.
7   Warren, *Brush with Death*, 66–67.
8   Markowitz and Rosner, *Deceit and Denial*.
9   Warren, *Brush with Death*.
10 Warren, 57.
11 Warren, 33.
12 Rabin, "The Lead Industry and Lead Water Pipes."
13 Marc Edwards, conversation with the author, October 17, 2016.

14 Sohn, "Lead."
15 Shavit and Shavit, "Lead and Arsenic in Morchella Esculenta Fruitbodies."
16 Ritter, "Pencils & Pencil Lead."
17 Van Ulirsch et al., "Evaluating and Regulating Lead in Synthetic Turf."
18 McGrayne, *Prometheans in the Lab*, 81.
19 "Selecting the Right Octane Fuel."
20 C. Williams, "Driving the Ford Model T through 110 Years of American Audacity."
21 Kitman, "The Secret History of Lead."
22 Kitman.
23 Rosner and Markowitz, "A 'Gift of God'?"
24 McGrayne, *Prometheans in the Lab*, 86.
25 McGrayne, 86–87.
26 Loeb, "Birth of the Kettering Doctrine."
27 McCarthy, *Auto Mania*, 48.
28 Warren, *Brush with Death*, 119–120.
29 McGrayne, *Prometheans in the Lab*, 89.
30 McGrayne, 90.
31 Rosner and Markowitz, "A 'Gift of God'?"
32 Rosner and Markowitz.
33 Kirk, Albin, and Elsma-Osorio, *The Impact of Environmental Law*.
34 Kovarik, "Ethyl-Leaded Gasoline."
35 Needleman, "Clamped in a Straitjacket."
36 McCarthy, *Auto Mania*, 49.
37 McGrayne, *Prometheans in the Lab*, 90
38 Quoted in Needleman, "Clamped in a Straitjacket."
39 Warren, *Brush with Death*, 121–122; McGrayne, *Prometheans in the Lab*, 90.
40 Kovarik, "Ethyl-Leaded Gasoline."
41 Rosner and Markowitz, "A 'Gift of God'?"
42 Needleman, "Clamped in a Straitjacket."
43 Rosner and Markowitz, "A 'Gift of God'?"
44 Rosner and Markowitz.
45 Rosner and Markowitz.
46 Rosner and Markowitz.
47 McGrayne, *Prometheans in the Lab*, 93.
48 Kovarik, "Ethyl-Leaded Gasoline."
49 Kitman, "The Secret History of Lead."
50 Kitman.
51 Gneezy, Meier, and Rey-Biel, "When and Why Incentives (Don't) Work to Modify Behavior."
52 Naughton, Sebold, and Mayer, "The Impacts of the California Beverage Container Recycling and Litter Reduction Act on Consumers."
53 Auten, Sieg, and Clotfelter, "Charitable Giving, Income, and Taxes."
54 Kovarik, "Ethyl-Leaded Gasoline."
55 U. Sinclair, *I, Candidate for Governor*, 109.
56 Rosner and Markowitz, "A 'Gift of God'?"
57 Rosner and Markowitz, *Dying for Work*.
58 Rosner and Markowitz, "A 'Gift of God'?"
59 Rosner and Markowitz.

60 Kovarik, "Ethyl-Leaded Gasoline."
61 Rosner and Markowitz, "A 'Gift of God'?"
62 Rosner and Markowitz.
63 McGrayne, *Prometheans in the Lab*, 94.
64 National Minerals Information Center, United States Geological Survey, "Historical Statistics for Mineral Commodities in the United States, Data Series 2005-140."

Chapter 3   Getting the Lead Out

1 Warren, *Brush with Death*, 28.
2 Advisory Committee on Childhood Lead Poisoning Prevention of the Centers for Disease Control and Prevention, "Low Level Lead Exposure Harms Children."
3 Gilbert and Weiss, "A Rationale for Lowering the Blood Lead Action Level from 10 to 2 µg/dL."
4 Bennett et al., "Project TENDR."
5 Agency for Toxic Substances and Disease Registry, U.S. Department of Health and Human Services, "Toxicological Profile for Lead."
6 Warren, *Brush with Death*, 42.
7 Warren, 129-130.
8 Warren, 130.
9 Gray, Cooke, and Tannenbaum, "Research Involving Human Subjects."
10 Warren, *Brush with Death*, 130-131; Denworth, *Toxic Truth*, 59-60.
11 Kehoe, Thamann, and Cholak, "On the Normal Absorption and Excretion of Lead."
12 Denworth, *Toxic Truth*, 59.
13 Denworth, 64-65.
14 Kitman, "The Secret History of Lead."
15 Denworth, *Toxic Truth*, 8-15.
16 Denworth, 21.
17 Denworth, 55-56.
18 Quoted in Denworth, 64.
19 Quoted in Denworth, 67.
20 Denworth, 47-76.
21 Denworth, 23-33.
22 Needleman, "Clamped in a Straitjacket."
23 Denworth, 77-80.
24 Needleman, Gunnoe, et al., "Deficits in Psychologic and Classroom Performance of Children with Elevated Dentine Lead Levels"; Needleman, Schell, et al., "Long-Term Effects of Exposure to Low Doses of Lead in Childhood."
25 Needleman, "Salem Comes to the National Institutes of Health."
26 Melnick, *Regulation and the Courts*, 5.
27 Mielke, "Dynamic Geochemistry of Tetraethyl Lead Dust during the 20th Century."
28 Howard Mielke, conversation with the author, July 26, 2016.
29 Mielke, conversation with the author, July 26, 2016.
30 Mielke, "Lead in the Inner Cities."
31 Warren, *Brush with Death*, 52.
32 Benfer, "Contaminated Childhood."
33 Kitman, "The Secret History of Lead."

34 National Minerals Information Center, United States Geological Survey, "Historical Statistics for Mineral Commodities in the United States, Data Series 2005-140."

35 Ludwig et al., "Survey of Lead in the Atmosphere of Three Urban Communities."

36 Quoted in Denworth, *Toxic Truth*, 74.

37 Quoted in Denworth, *Toxic Truth*, 75–76.

38 US EPA, "History of Reducing Air Pollution from Transportation in the United States."

39 US EPA.

40 Newell and Rogers, "The U.S. Experience with the Phasedown of Lead in Gasoline."

41 Melnick, *Regulation and the Courts*, 241.

42 National Minerals Information Center, United States Geological Survey, "Historical Statistics for Mineral Commodities in the United States, Data Series 2005-140."

43 Kitman, "The Secret History of Lead."

44 Quoted in Denworth, *Toxic Truth*, 95.

45 Kitman, "The Secret History of Lead."

46 Trudeau, *Doonesbury*, January 28, 1982.

47 Dennis and Mooney, "Neil Gorsuch's Mother Once Ran the EPA."

48 Markowitz and Rosner, *Lead Wars*, 190.

49 Markowitz and Rosner, 110.

50 Markowitz and Rosner, 111.

51 Morgenstern, *Economic Analyses at EPA*.

52 Schwartz et al., "Costs and Benefits of Reducing Lead in Gasoline."

53 Hays, "The Role of Values in Science and Policy," 305.

54 Newell and Rogers, "The U.S. Experience with the Phasedown of Lead in Gasoline."

55 National Minerals Information Center, United States Geological Survey, "Historical Statistics for Mineral Commodities in the United States, Data Series 2005-140."

56 US EPA, "EPA Takes Final Step in Phaseout of Leaded Gasoline."

57 "Selecting the Right Octane Fuel."

Chapter 4   Lead in America's Children

1   Markowitz and Rosner, *Lead Wars*, 31.

2   Warren, *Brush with Death*, 163; "Huntington Williams, Was Health Commissioner."

3   H. Williams et al., "Lead Poisoning in Young Children."

4   Berney, "Round and Round It Goes."

5   Fine et al., "Pediatric Blood Lead Levels."

6   Robbins et al., "Childhood Lead Exposure and Uptake in Teeth in the Cleveland Area."

7   Robbins et al.

8   Robbins et al.

9   Warren, *Brush with Death*, 211.

10 Annest and Mahaffey, *Blood Lead Levels for Persons Ages 6 Months–74 Years*.

11 Annest and Mahaffey.

12 Pirkle et al., "The Decline in Blood Lead Levels in the United States."
13 Robbins et al., "Childhood Lead Exposure and Uptake in Teeth in the Cleveland Area."
14 McNeill, *Something New under the Sun*, 62.
15 Centers for Disease Control and Prevention, "Blood Lead Levels in Children Aged 1–5 Years—United States, 1999–2010."
16 Annest and Mahaffey, *Blood Lead Levels for Persons Ages 6 Months–74 Years*.
17 Pirkle et al., "The Decline in Blood Lead Levels in the United States."
18 Centers for Disease Control and Prevention, "Blood Lead Levels in Children Aged 1–5 Years—United States, 1999–2010."
19 Hillier, "Redlining and the Home Owners' Loan Corporation."
20 Hillier.
21 Hillier.
22 Lambert, "Study Calls LI Most Segregated Suburb."
23 Kranish, "Decades-Old Housing Discrimination Case Plagues Donald Trump."
24 Hogan and Berry, "Racial and Ethnic Biases in Rental Housing."
25 Bayer et al., "Racial and Ethnic Price Differentials in the Housing Market."
26 Bruenig, "The Racial Wealth Gap."
27 Ostrander, "School Funding."
28 Morland et al., "Neighborhood Environment and Adiposity among Older Adults."
29 Roosevelt, "Toll Roads and Free Roads."
30 Connerly, "From Racial Zoning to Community Empowerment."
31 Little and Wiffen, "Emission and Deposition of Lead from Motor Exhausts—II."
32 Butterworth and Hopkins, "Hand-Mouth Coordination in the New-Born Baby"; Kurjak et al., "Fetal Hand Movements and Facial Expression in Normal Pregnancy."

Chapter 5   Brains and Behavior and Lead

1   Stiles and Jernigan, "The Basics of Brain Development."
2   Graham and Forstadt, "Children and Brain Development."
3   Graham and Forstadt.
4   Deoni et al., "Mapping Infant Brain Myelination with Magnetic Resonance Imaging"; Snaidero and Simons, "Myelination at a Glance."
5   Innocenti and Price, "Exuberance in the Development of Cortical Networks."
6   Heizman, "Calcium Signaling in the Brain."
7   Bressler and Goldstein, "Mechanisms of Lead Neurotoxicity."
8   Bressler and Goldstein.
9   Sanders et al., "Neurotoxic Effects and Biomarkers of Lead Exposure."
10 Advisory Committee on Childhood Lead Poisoning Prevention of the Centers for Disease Control and Prevention, "Low Level Lead Exposure Harms Children."
11 Lanphear, Rauch, et al., "Low-Level Lead Exposure and Mortality in US Adults."
12 Ekong, Jaar, and Weaver, "Lead-Related Nephrotoxicity."
13 Ekong, Jaar, and Weaver.
14 Winder, "Lead, Reproduction and Development."
15 Altmann et al., "Impairment of Long-Term Potentiation and Learning following Chronic Lead Exposure."
16 Altmann et al.
17 Cecil et al., "Decreased Brain Volume in Adults with Childhood Lead Exposure."

18 Cecil et al.
19 Miller and Cohen, "An Integrative Theory of Prefrontal Cortex Function."
20 Landrigan et al., "Neuropsychological Dysfunction in Children with Chronic Low-Level Lead Absorption."
21 Needleman, Gunnoe, et al., "Deficits in Psychologic and Classroom Performance of Children with Elevated Dentine Lead Levels."
22 Rosselli and Ardila, "The Impact of Culture and Education on Non-verbal Neuropsychological Measurements."
23 Gardner and Hatch, "Multiple Intelligences Go to School."
24 Paransky and Zurawin, "Management of Menstrual Problems and Contraception in Adolescents with Mental Retardation."
25 Herrnstein and Murray, The Bell Curve.
26 Jensen and Nyborg, The Scientific Study of General Intelligence.
27 Lanphear, Hornung, et al., "Low-Level Environmental Lead Exposure and Children's Intellectual Function."
28 Reyes, "Lead Policy and Academic Performance."
29 Miranda et al., "The Relationship between Early Childhood Blood Lead Levels and Performance on End-of-Grade Tests."
30 Amato et al., "Lead Exposure and Educational Proficiency."
31 Chandramouli et al., "Effects of Early Childhood Lead Exposure on Academic Performance and Behaviour of School Age Children."
32 Wang et al., "Relationship between Blood Lead Concentrations and Learning Achievement among Primary School Children in Taiwan."
33 Kordas et al., "Deficits in Cognitive Function and Achievement in Mexican First-Graders with Low Blood Lead Concentrations."
34 Reuben et al., "Association of Childhood Blood Lead Levels with Cognitive Function and Socioeconomic Status."
35 Reuben et al.
36 Needleman, Gunnoe, et al., "Deficits in Psychologic and Classroom Performance of Children with Elevated Dentine Lead Levels."
37 Nigg et al., "Confirmation and Extension of Association of Blood Lead with Attention-Deficit/Hyperactivity Disorder."
38 Nigg et al.; Goodlad, Marcus, and Fulton, "Lead and Attention-Deficit/Hyperactivity Disorder (ADHD) Symptoms."
39 Mendelsohn et al., "Low-Level Lead Exposure and Behavior in Early Childhood."
40 Tuthill, "Hair Lead Levels Related to Children's Classroom Attention-Deficit Behavior."
41 Braun et al., "Association of Environmental Toxicants and Conduct Disorder in U.S. Children."
42 Needleman, Riess, et al., "Bone Lead Levels and Delinquent Behavior."
43 Marcus, Fulton, and Clarke, "Lead and Conduct Problems."
44 Katsiyannis et al., "Juvenile Delinquency and Recidivism"; Krakowski, "Violence and Serotonin."

Chapter 6 Lead and Violence

1 Delville, "Exposure to Lead during Development Alters Aggressive Behavior in Golden Hamsters."
2 Li et al., "Lead Exposure Potentiates Predatory Attack Behavior in the Cat."

3 Baum, "Legalize It All."
4 Chokshi, "Black People More Likely to Be Wrongfully Convicted of Murder, Study Shows"; Winerip, Schwirtz, and Gebeloff, "For Blacks Facing Parole in New York State, Signs of a Broken System"; Anwar, Bayer, and Hjalmarsson, "The Impact of Jury Race in Criminal Trials."
5 Zimring and Hawkins, *Crime Is Not the Problem*, 102.
6 National Research Council of the National Academies, *Understanding Crime Trends*.
7 National Research Council of the National Academies, 164.
8 Tcherni-Buzzeo, "The 'Great American Crime Decline.'"
9 Needleman, McFarland, et al., "Bone Lead Levels in Adjudicated Delinquents."
10 Needleman, McFarland, et al.
11 Dietrich et al., "Early Exposure to Lead and Juvenile Delinquency."
12 Dietrich et al., "Early Exposure to Lead and Juvenile Delinquency," 513.
13 Wright et al., "Association of Prenatal and Childhood Blood Lead Concentrations with Criminal Arrests in Early Adulthood."
14 Wright et al.
15 Grönqvist, Nilsson, and Robling, "Early Lead Exposure and Outcomes in Adulthood."
16 Grönqvist, Nilsson, and Robling.
17 "Bureau of Justice Statistics (BJS)—National Crime Victimization Survey (NCVS)."
18 Federal Bureau of Investigation, "Uniform Crime Reporting Statistics."
19 Gallup, Inc., "More Americans Say Crime Is Rising in U.S."
20 Garofalo, "NCCD Research Review."
21 Pinker, *The Better Angels of Our Nature*.
22 Nevin, "How Lead Exposure Relates to Temporal Changes in IQ, Violent Crime, and Unwed Pregnancy."
23 Nevin.
24 Drum, "Lead: America's Real Criminal Element."
25 Vigen, *Spurious Correlations*.
26 Reyes, "Environmental Policy as Social Policy?"
27 Reyes.
28 Mielke and Zahran, "The Urban Rise and Fall of Air Lead (Pb) and the Latent Surge and Retreat of Societal Violence."
29 Boutwell et al., "The Intersection of Aggregate-Level Lead Exposure and Crime."
30 Nevin, "Understanding International Crime Trends."
31 McNeill, *Something New under the Sun*, 62.
32 Zimring and Hawkins, *Crime Is Not the Problem*, xi.
33 Lovei, "Phasing Out Lead from Gasoline."
34 Nevin, *Lucifer Curves*, 63.
35 Zimring and Hawkins, *Crime Is Not the Problem*, 106.
36 Taylor et al., "The Relationship between Atmospheric Lead Emissions and Aggressive Crime."
37 Masters, Hone, and Doshi, "Environmental Pollution, Neurotoxicity, and Criminal Violence."
38 Feigenbaum and Muller, "Lead Exposure and Violent Crime in the Early Twentieth Century."
39 Feigenbaum and Muller.

40 Lauritsen, Rezey, and Heimer, "When Choice of Data Matters."

41 Lauritsen, Rezey, and Heimer.

42 Kindermann, Lynch, and Cantor, "Effects of the Redesign on Victimization Estimates."

43 Cory-Slechta et al., "Maternal Stress Modulates the Effects of Developmental Lead Exposure"; Bellinger et al., "Low-Level Lead Exposure, Social Class, and Infant Development."

44 Cory-Slechta et al., "Maternal Stress Modulates the Effects of Developmental Lead Exposure."

45 Guilarte et al., "Environmental Enrichment Reverses Cognitive and Molecular Deficits Induced by Developmental Lead Exposure."

46 Lidsky and Schneider, "Lead Neurotoxicity in Children."

47 Fazel et al., "Risk of Violent Crime in Individuals with Epilepsy and Traumatic Brain Injury."

48 Zimring and Hawkins, *Crime Is Not the Problem*, 101.

49 Zimring and Hawkins, 189.

50 Zimring and Hawkins, 99.

51 Rhoten, "Risks and Rewards of an Interdisciplinary Research Path."

52 İmrohoroğlu, Merlo, and Rupert, "What Accounts for the Decline in Crime?"; Levitt, "Understanding Why Crime Fell in the 1990s"; Donohue and Levitt, "The Impact of Legalized Abortion on Crime"; Sharkey, Torrats-Espinosa, and Takyar, "Community and the Crime Decline."

53 Bernhoft, "Mercury Toxicity and Treatment"; Rodríguez-Barranco et al., "Association of Arsenic, Cadmium and Manganese Exposure with Neurodevelopment and Behavioural Disorders in Children"; Iguchi, Watanabe, and Katsu, "Developmental Effects of Estrogenic Agents on Mice, Fish, and Frogs"; Oberdörster, Elder, and Rinderknecht, "Nanoparticles and the Brain."

54 Zimring and Hawkins, *Crime Is Not the Problem*, 190.

55 Katz, "Environmental Policy Guide."

56 Raine, *The Anatomy of Violence*, 11–13.

57 Lafree, Baumer, and O'Brien, "Still Separate and Unequal?"

58 Matthews, "Here's What You Need to Know about Stop and Frisk—and Why the Courts Shut It Down."

59 "Bureau of Justice Statistics (BJS)—Traffic Stops."

60 Lowery, "Aren't More White People Than Black People Killed by Police?"

61 "Report: The War on Marijuana in Black and White."

62 Rehavi and Starr, "Racial Disparity in Federal Criminal Sentences."

63 Alexander and West, *The New Jim Crow*.

64 Chokshi, "Black People More Likely to Be Wrongfully Convicted of Murder, Study Shows"; Anwar, Bayer, and Hjalmarsson, "The Impact of Jury Race in Criminal Trials"; Winerip, Schwirtz, and Gebeloff, "For Blacks Facing Parole in New York State, Signs of a Broken System."

65 Payne, Vuletich, and Lundberg, "The Bias of Crowds."

66 Rachlinski et al., "Does Unconscious Racial Bias Affect Trial Judges?"

67 Douglas et al., "Risk Assessment Tools in Criminal Justice and Forensic Psychiatry."

68 Israni, "When an Algorithm Helps Send You to Prison."

69 Angwin et al., "Machine Bias."

70 Dilulio, "The Coming of the Super-Predators."

71 Federal Bureau of Investigation, "Uniform Crime Reporting Statistics."
72 "Bureau of Justice Statistics (BJS)—National Crime Victimization Survey (NCVS)."
73 Fox, "Trends in Juvenile Violence."
74 Snyder, "Arrest in the United States, 1990–2010."
75 Glaeser and Sacerdote, "Why Is There More Crime in Cities?"
76 Myers et al., "Safety in Numbers."
77 J. Goodman and Baker, "Murders in New York Drop to a Record Low, but Officers Aren't Celebrating."
78 Zimring, *The Great American Crime Decline*, 106.
79 New York State Department of Labor, "Labor Statistics for the New York City Region"
80 Carver, Timperio, and Crawford, "Playing It Safe."
81 Lahey, *The Gift of Failure*.
82 Lythcott-Haims, *How to Raise an Adult*.
83 *Free Range Kids* (blog).

Chapter 7   The Lead Problem Persists

1 Gilbert and Weiss, "A Rationale for Lowering the Blood Lead Action Level from 10 to 2 µg/dL."
2 Hanna-Attisha, *What the Eyes Don't See*, 117–130.
3 Masten, Davies, and Mcelmurry, "Flint Water Crisis."
4 Masten, Davies, and Mcelmurry.
5 Kennedy, "Lead-Laced Water in Flint."
6 Roy, "Our Sampling of 252 Homes Demonstrates a High Lead in Water Risk."
7 Masten, Davies, and Mcelmurry, "Flint Water Crisis."
8 Hanna-Attisha et al., "Elevated Blood Lead Levels in Children Associated with the Flint Drinking Water Crisis."
9 Vaccari, "How Not to Get the Lead Out."
10 Michigan Civil Rights Commission, "The Flint Water Crisis."
11 Quoted in Michigan Civil Rights Commission, "The Flint Water Crisis," 2.
12 Warren, *Brush with Death*, 33.
13 Jacobs et al., "The Prevalence of Lead-Based Paint Hazards in U.S. Housing."
14 Markowitz, "The Childhood Lead Poisoning Epidemic in Historical Perspective."
15 "Lead and Healthy Homes Program."
16 "Lead and Healthy Homes Program."
17 Centers for Disease Control and Prevention, "Blood Lead Levels in Children Aged 1–5 Years—United States, 1999–2010."
18 US EPA, "Renovation, Repair and Painting Program."
19 US EPA.
20 Markowitz and Rosner, *Lead Wars*.
21 Markowitz and Rosner.
22 Markowitz and Rosner.
23 Philadelphia Childhood Lead Poisoning Prevention Advisory Group, "Final Report and Recommendations."
24 US EPA, "Real Estate Disclosure."
25 Hurwitz and Peffley, "Explaining the Great Racial Divide."
26 McCoy, "Freddie Gray's Life a Study on the Effects of Lead Paint on Poor Blacks."

27 Markowitz and Rosner, *Lead Wars*.
28 Hiltzik, "Supreme Court Deals Final Blow to Lead Paint Manufacturers' Years-Long Effort to Avoid Cleanup Costs."
29 Dwyer, "Michigan's Office Reviewing Lawsuit to Hold Companies Liable for Putting Lead in Products."
30 Howard Mielke, conversation with the author, July 26, 2016.
31 Mielke et al., "Nonlinear Association between Soil Lead and Blood Lead of Children in Metropolitan New Orleans, Louisiana."
32 US EPA, OECA, "Q&A."
33 Ruderman, Laker, and Purcell, "In Booming Philly Neighborhoods, Lead-Poisoned Soil Is Resurfacing."
34 National Center for Environmental Health, "CDC—Lead—Tips—Sources of Lead—Toy Jewelry."
35 National Center for Environmental Health, "CDC—Lead—Tips—Sources of Lead—Sindoor Alert."
36 Van Ulirsch et al., "Evaluating and Regulating Lead in Synthetic Turf."
37 Health Impact Project, "10 Policies to Prevent and Respond to Childhood Lead Exposure," 63–64.
38 Federal Aviation Administration, "Aviation Gasoline."
39 Gould, "Childhood Lead Poisoning."
40 Health Impact Project, "10 Policies to Prevent and Respond to Childhood Lead Exposure," 2.
41 Health Impact Project, 2.

Chapter 8 Lessons from the Lead Battles

1   *The Global Risks Report 2018*.
2   Chan, "Coal Was Used as Fuel in China More Than 3,500 Years Ago."
3   Maryon, "Metal Working in the Ancient World"; Condliffe, "Lost Treasures"; Kurlansky, "'Salt.'"
4   Intergovernmental Panel on Climate Change, "Working Group II: Impacts, Adaptation and Vulnerability"
5   Funk and Kennedy, "Public Confidence in Scientists Has Remained Stable for Decades."
6   B. King, "Pop Quiz."
7   Needleman, "Salem Comes to the National Institutes of Health."
8   Needleman.
9   "Historical Sea Surface Temperature Adjustments/Corrections aka 'The Bucket Model.'"
10 Oreskes and Conway, *Merchants of Doubt*.
11 Cornwall, "New Rule Could Force EPA to Ignore Major Human Health Studies."
12 "Climate Oscillations and the Global Warming Faux Pause."
13 Warren, *Brush with Death*.
14 US EPA, "Light-Duty Vehicle Greenhouse Gas Emission Standards."
15 "Process Matters."
16 Oreskes and Conway, *Merchants of Doubt*.
17 Schneider, *Science as a Contact Sport*.
18 B. King, "Pop Quiz."
19 Kolbert, "Why Facts Don't Change Our Minds."

20 Oreskes and Conway, *Merchants of Doubt.*
21 Hays, "The Role of Values in Science and Policy."
22 Oreskes and Conway, *Merchants of Doubt.*
23 "NASA Sea Level Change Portal."
24 Needleman, "The Removal of Lead from Gasoline."
25 Needleman.
26 Hershey, "How the Oil Glut Is Changing Business."
27 Newell and Rogers, "The U.S. Experience with the Phasedown of Lead in Gasoline."
28 "How Much Will the Clean Power Plan Cost?"
29 "Obama's 'Clean Power Plan' Should Be Called the 'Costly Power Plan.'"
30 Ross, Hoppock, and Murray, "Ongoing Evolution of the Electricity Industry."
31 Grab and Lienke, "The Falling Cost of Clean Power Plan Compliance."
32 Tonachel, "2025 Clean Car Standards Are Achievable, Study Shows."
33 DeCicco, "The 'Job-Killing' Fiction behind Trump's Retreat on Fuel Economy Standards."
34 Plungis, "EPA Chief Rejects Obama-Era Fuel Economy Targets."
35 Furth, "Fuel Economy Standards Are a Costly Mistake."
36 "Bureau of Labor Statistics Data."
37 Jaffe et al., "Environmental Regulation and the Competitiveness of U.S. Manufacturing."
38 Yang, "Does Government Regulation Really Kill Jobs?"
39 T. Sinclair and Vesey, "Regulation, Jobs, and Economic Growth."
40 Goldschlag and Tabarrok, "Is Regulation to Blame for the Decline in American Entrepreneurship?"
41 Cohen, "The Libertarian Who Accidentally Helped Make the Case for Regulation."
42 Goldschlag and Tabarrok, "Is Regulation to Blame for the Decline in American Entrepreneurship?"
43 Blumberg, "First Direct Proof of Ozone Hole Recovery Due to Chemicals Ban."
44 "When Our Rivers Caught Fire."
45 US Census Bureau, "Estimates of U.S. Population by Age and Sex."
46 "Our Nation's River."
47 US EPA, "Air Quality—National Summary."
48 Weiss, "Is Acid Rain a Thing of the Past?"
49 Flanagan, "Asbestos Data Sheet."
50 Bialik, "Most Americans Favor Stricter Environmental Laws and Regulations."
51 Bialik.
52 Hays, "The Role of Values in Science and Policy," 309.
53 Henry, "US R&D Spending at All-Time High, Federal Share Reaches Record Low."
54 Scialla, "It Could Take Centuries for EPA to Test All the Unregulated Chemicals under a New Landmark Bill."
55 Hoffman, Sosa, and Stapleton, "Do Flame Retardant Chemicals Increase the Risk for Thyroid Dysregulation and Cancer?"; Gao et al., "Bisphenol A and Hormone-Associated Cancers."
56 Frankel, "The Government Won't Fund Research on Gun Violence Because of NRA Lobbying."
57 Greenfield Boyce, "Spending Bill Lets CDC Study Gun Violence."
58 Holodny, "Traffic Fatalities in the US Have Been Mostly Plummeting for Decades."

59 Melnick, *Regulation and the Courts.*
60 Freed and Currinder, "Do Political Business in the Daylight"; Drutman, "How Corporate Lobbyists Conquered American Democracy."
61 Overby, "Once Ruled by Washington Insiders, Campaign Finance Reform Goes Grass Roots."
62 Needleman, "The Removal of Lead from Gasoline."
63 Hays, "The Role of Values in Science and Policy," 303.
64 Mooney and Dennis, "On Climate Change, Scott Pruitt Causes an Uproar."
65 Meko, Lu, and Gamio, "How Trump Won the Presidency with Razor-Thin Margins in Swing States."
66 Weisburd et al., "Understanding the Mechanisms Underlying Broken Windows Policing."
67 "Stop-and-Frisk Data."
68 Zimring, Kamin, and Hawkins, *Crime and Punishment in California.*
69 Taibbi, "The Shame of Three Strikes Laws."
70 "Criminal Justice Facts."
71 Lee, "Does the United States Really Have 5 Percent of the World's Population and One Quarter of the World's Prisoners?"
72 E. King, "Black Men Get Longer Prison Sentences Than White Men for the Same Crime."
73 Picchi, "The High Price of Incarceration in America."
74 Chadwick, *Theodore Parker.*
75 Parker, "Of Justice and the Conscience."

Conclusion

1 Warren, *Brush with Death,*125–127; Kitman, "The Secret History of Lead."
2 Lovei, "Phasing Out Lead from Gasoline."
3 Eveleth, "Barns Are Painted Red Because of the Physics of Dying Stars."
4 McGrayne, *Prometheans in the Lab*, 81.
5 Warren, *Brush with Death*, 178.
6 Schwartz et al., "Costs and Benefits of Reducing Lead in Gasoline."
7 Newell and Rogers, "The U.S. Experience with the Phasedown of Lead in Gasoline."
8 Melnick, *Regulation and the Courts*, 5.
9 Sharkey, *Uneasy Peace*, 60.
10 Sharkey, 57.
11 Roeder, Eisen, and Bowling, "What Caused the Crime Decline?," 62–63.
12 Zimring and Hawkins, *Crime Is Not the Problem*, 99.
13 Zimring and Hawkins, 103.

Bibliography

Advisory Committee on Childhood Lead Poisoning Prevention of the Centers for Disease Control and Prevention. "Low Level Lead Exposure Harms Children: A Renewed Call for Primary Prevention," January 4, 2012. https://www.cdc.gov /nceh/lead/acclpp/final_document_030712.pdf.

Agency for Toxic Substances and Disease Registry, U.S. Department of Health and Human Services. "Toxicological Profile for Lead," May 2019. https://www.atsdr .cdc.gov/toxprofiles/tp13.pdf.

Alexander, Michelle, and Cornel West. *The New Jim Crow: Mass Incarceration in the Age of Colorblindness*. Rev. ed. New York: New Press, 2012.

Altmann, Lilo, Frank Weinsberg, Karolina Sveinsson, Hellmuth Lilienthal, Herbert Wiegand, and Gerhard Winneke. "Impairment of Long-Term Potentiation and Learning following Chronic Lead Exposure." *Toxicology Letters* 66, no. 1 (January 1, 1993): 105–112. https://doi.org/10.1016/0378-4274(93)90085-C.

Amato, Michael S., Colleen F. Moore, Sheryl Magzamen, Pamela Imm, Jeffrey A. Havlena, Henry A. Anderson, and Marty S. Kanarek. "Lead Exposure and Educational Proficiency: Moderate Lead Exposure and Educational Proficiency on End-of-Grade Examinations." *Annals of Epidemiology* 22, no. 10 (October 1, 2012): 738–743. https://doi.org/10.1016/j.annepidem.2012.07.004.

Angwin, Julia, Jeff Larson, Surya Mattu, and Lauren Kirchner. "Machine Bias." ProPublica, May 23, 2016. https://www.propublica.org/article/machine-bias-risk -assessments-in-criminal-sentencing.

Annest, Joseph L., and Kathryn R. Mahaffey. *Blood Lead Levels for Persons Ages 6 Months–74 Years: United States, 1976–80*. DHHS Publication No. 84-1683. Hyattsville, MD: U.S. Dept. of Health and Human Services, Public Health Service, National Center for Health Statistics, 1984.

Anwar, Shamena, Patrick Bayer, and Randi Hjalmarsson. "The Impact of Jury Race in Criminal Trials." *Quarterly Journal of Economics* 127, no. 2 (May 1, 2012): 1017–1055. https://doi.org/10.1093/qje/qjs014.

Auten, Gerald E., Holger Sieg, and Charles T. Clotfelter. "Charitable Giving, Income, and Taxes: An Analysis of Panel Data." *American Economic Review* 92, no. 1 (2002): 371–382.

Baum, Dan. "Legalize It All." *Harper's Magazine*, April 2016. https://harpers.org /archive/2016/04/legalize-it-all/.

Bayer, Patrick, Marcus Casey, Fernando Ferreira, and Robert McMillan. "Racial and Ethnic Price Differentials in the Housing Market." *Journal of Urban Economics* 102 (November 2017): 91–105. https://doi.org/10.1016/j.jue.2017.07.004.

Bellinger, David, Alan Leviton, Christine Waternaux, Herbert Needleman, and Michael Rabinowitz. "Low-Level Lead Exposure, Social Class, and Infant Development." *Neurotoxicology and Teratology* 10, no. 6 (November 1, 1988): 497–503. https://doi.org/10.1016/0892-0362(88)90084-0.

Benfer, Emily A. "Contaminated Childhood: How the United States Failed to Prevent the Chronic Lead Poisoning of Low-Income Children and Communities of Color." *Harvard Environmental Law Review* 41, no. 2 (2017): 493–561.

Bennett, Deborah, David C. Bellinger, Linda S. Birnbaum, Asa Bradman, Aimin Chen, Deborah A. Cory-Slechta, Stephanie M. Engel, et al. "Project TENDR: Targeting Environmental Neuro-Developmental Risks: The TENDR Consensus Statement." *Environmental Health Perspectives* 124, no. 7 (July 1, 2016): A118–122. https://doi.org/10.1289/EHP358.

Berners-Lee, Mike. *How Bad Are Bananas? The Carbon Footprint of Everything.* Vancouver: Greystone Books, 2011.

Berney, Barbara. "Round and Round It Goes: The Epidemiology of Childhood Lead Poisoning, 1950–1990." *Milbank Quarterly* 71, no. 1 (1993): 3–39. https://doi.org/10 .2307/3350273.

Bernhoft, Robin A. "Mercury Toxicity and Treatment: A Review of the Literature." *Journal of Environmental and Public Health* 2012, (2012). https://doi.org/10.1155 /2012/460508.

Bialik, Kristen. "Most Americans Favor Stricter Environmental Laws and Regulations." *Pew Research Center* (blog), December 14, 2016. http://www.pewresearch .org/fact-tank/2016/12/14/most-americans-favor-stricter-environmental-laws-and -regulations/.

Blumberg, Sara. "First Direct Proof of Ozone Hole Recovery Due to Chemicals Ban." NASA, January 4, 2018. http://www.nasa.gov/feature/goddard/2018/nasa-study -first-direct-proof-of-ozone-hole-recovery-due-to-chemicals-ban.

Boutwell, Brian B., Erik J. Nelson, Brett Emo, Michael G. Vaughn, Mario Schootman, Richard Rosenfeld, and Roger Lewis. "The Intersection of Aggregate-Level Lead Exposure and Crime." *Environmental Research* 148 (July 2016): 79–85. https://doi .org/10.1016/j.envres.2016.03.023.

Braun, Joseph M., Tanya E. Froehlich, Julie L. Daniels, Kim N. Dietrich, Richard Hornung, Peggy Auinger, and Bruce P. Lanphear. "Association of Environmental Toxicants and Conduct Disorder in U.S. Children: NHANES 2001–2004." *Environmental Health Perspectives* 116, no. 7 (July 2008): 956–962. https://doi.org /10.1289/ehp.11177.

Bressler, Joseph P., and Gary W. Goldstein. "Mechanisms of Lead Neurotoxicity." *Biochemical Pharmacology* 41, no. 4 (February 15, 1991): 479–484. https://doi.org /10.1016/0006-2952(91)90617-E.

Bruenig, Matt. "The Racial Wealth Gap." *Demos*, November 5, 2013. Accessed August 15, 2018. https://www.demos.org/blog/11/5/13/racial-wealth-gap.

"Bureau of Justice Statistics (BJS)—National Crime Victimization Survey (NCVS)." Accessed August 15, 2018. https://www.bjs.gov/index.cfm?ty=dcdetail&iid =245.

"Bureau of Justice Statistics (BJS)—Traffic Stops." Accessed August 16, 2018. https://www.bjs.gov/index.cfm?tid=702&ty=tp.

"Bureau of Labor Statistics Data." Accessed August 16, 2018. https://data.bls.gov/timeseries/CES3133600101?amp%253bdata_tool=XGtable&output_view=data&include_graphs=true.

Butterworth, George, and Brian Hopkins. "Hand-Mouth Coordination in the New-Born Baby." *British Journal of Developmental Psychology* 6, no. 4 (November 1, 1988): 303–314. https://doi.org/10.1111/j.2044-835X.1988.tb01103.x.

Carver, Alison, Anna Timperio, and David Crawford. "Playing It Safe: The Influence of Neighbourhood Safety on Children's Physical Activity—A Review." *Health and Place* 14, no. 2 (June 1, 2008): 217–227. https://doi.org/10.1016/j.healthplace.2007.06.004.

Cecil, Kim M., Christopher J. Brubaker, Caleb M. Adler, Kim N. Dietrich, Mekibib Altaye, John C. Egelhoff, Stephanie Wessel, et al. "Decreased Brain Volume in Adults with Childhood Lead Exposure." *PLOS Medicine* 5, no. 5 (May 27, 2008): e112. https://doi.org/10.1371/journal.pmed.0050112.

Centers for Disease Control and Prevention. "Blood Lead Levels in Children Aged 1–5 Years—United States, 1999–2010." *Morbidity and Mortality Weekly Report* 62, no. 13 (April 5, 2013): 245–248. https://www.cdc.gov/mmwr/preview/mmwrhtml/mm6213a3.htm.

Chadwick, John White. *Theodore Parker: Preacher and Reformer.* London: Forgotten Books, 2015.

Chan, Emily. "Coal Was Used as Fuel in China More Than 3,500 Years Ago." *Daily Mail*, August 18, 2015. Accessed August 16, 2018. http://www.dailymail.co.uk/news/peoplesdaily/article-3202104/Remarkable-discovery-shows-humans-burning-coal-fuel-3-500-years-ago-world-s-earliest-site-activity-unearthed-China.html.

Chandramouli, K., C. D. Steer, M. Ellis, and A. M. Emond. "Effects of Early Childhood Lead Exposure on Academic Performance and Behaviour of School Age Children." *Archives of Disease in Childhood* 94, no. 11 (November 1, 2009): 844–848. https://doi.org/10.1136/adc.2008.149955.

Chokshi, Niraj. "Black People More Likely to Be Wrongfully Convicted of Murder, Study Shows." *New York Times*, January 20, 2018, U.S. sec. https://www.nytimes.com/2017/03/07/us/wrongful-convictions-race-exoneration.html.

"Climate Oscillations and the Global Warming Faux Pause." *RealClimate* (blog), February 26, 2015. Accessed August 16, 2018. http://www.realclimate.org/index.php/archives/2015/02/climate-oscillations-and-the-global-warming-faux-pause/.

Cohen, Rachel. "The Libertarian Who Accidentally Helped Make the Case for Regulation." *Washington Monthly*, April 8, 2018. https://washingtonmonthly.com/magazine/april-may-june-2018/null-hypothesis/.

Condliffe, Jamie. "Lost Treasures: The Napalm of Byzantium." *New Scientist*, February 1, 2012. Accessed August 16, 2018. https://www.newscientist.com/article/mg21328502-400-lost-treasures-the-napalm-of-byzantium/.

Connerly, Charles E. "From Racial Zoning to Community Empowerment: The Interstate Highway System and the African American Community in Birmingham, Alabama." *Journal of Planning Education and Research* 22, no. 2 (December 1, 2002): 99–114. https://doi.org/10.1177/0739456X02238441.

Cornwall, Warren. "New Rule Could Force EPA to Ignore Major Human Health Studies." *Science*, April 25, 2018. http://www.sciencemag.org/news/2018/04/new-rule-could-force-epa-ignore-major-human-health-studies.

Cory-Slechta, Deborah A., Miriam B. Virgolini, Mona Thiruchelvam, Doug D. Weston, and Mark R. Bauter. "Maternal Stress Modulates the Effects of Developmental Lead Exposure." *Environmental Health Perspectives* 112, no. 6 (May 2004): 717–730. https://doi.org/10.1289/ehp.6481.

"Criminal Justice Facts." The Sentencing Project. Accessed August 16, 2018. https://www.sentencingproject.org/criminal-justice-facts/.

DeCicco, John. "The 'Job-Killing' Fiction behind Trump's Retreat on Fuel Economy Standards." *Yale Environment 360*, March 20, 2017. Accessed August 16, 2018. https://e360.yale.edu/features/trump-fuel-economy-cafe-standards-decicco.

Delile, Hugo, Janne Blichert-Toft, Jean-Philippe Goiran, Simon Keay, and Francis Albarède. "Lead in Ancient Rome's City Waters." *Proceedings of the National Academy of Sciences* 111, no. 18 (May 6, 2014): 6594–6599. https://doi.org/10.1073/pnas.1400097111.

Delville, Yvon. "Exposure to Lead during Development Alters Aggressive Behavior in Golden Hamsters." *Neurotoxicology and Teratology* 21, no. 4 (July 1, 1999): 445–449. https://doi.org/10.1016/S0892-0362(98)00062-2.

Dennis, Brady, and Chris Mooney, "Neil Gorsuch's Mother Once Ran the EPA. It Didn't Go Well." *Washington Post*, February 1, 2017. Accessed August 15, 2018. https://www.washingtonpost.com/news/energy-environment/wp/2017/02/01/neil-gorsuchs-mother-once-ran-the-epa-it-was-a-disaster/?utm_term=.f0750e17bc67.

Denworth, Lydia. *Toxic Truth: A Scientist, a Doctor, and the Battle over Lead*. Boston: Beacon Press, 2008.

Deoni, Sean C. L., Evelyne Mercure, Anna Blasi, David Gasston, Alex Thomson, Mark Johnson, Steven C. R. Williams, and Declan G. M. Murphy. "Mapping Infant Brain Myelination with Magnetic Resonance Imaging." *Journal of Neuroscience* 31, no. 2 (January 12, 2011): 784–791. https://doi.org/10.1523/JNEUROSCI.2106-10.2011.

Dietrich, Kim N., Ris M. Douglas, Paul A. Succop, Omer G. Berger, and Robert L. Bornschein. "Early Exposure to Lead and Juvenile Delinquency." *Neurotoxicology and Teratology* 23, no. 6 (2001): 511–518.

Dilulio, John. "The Coming of the Super-Predators." *Weekly Standard*, November 27, 1995. Accessed August 16, 2018. https://www.weeklystandard.com/the-coming-of-the-super-predators/article/8160.

Dodds, Walter K., and Val H. Smith. "Nitrogen, Phosphorus, and Eutrophication in Streams." *Inland Waters* 6, no. 2 (January 1, 2016): 155–164. https://doi.org/10.5268/IW-6.2.909.

Donohue, John J., and Steven D. Levitt. "The Impact of Legalized Abortion on Crime." *Quarterly Journal of Economics* 116, no. 2 (May 1, 2001): 379–420. https://doi.org/10.1162/00335530151144050.

Douglas, T., J. Pugh, I. Singh, J. Savulescu, and S. Fazel. "Risk Assessment Tools in Criminal Justice and Forensic Psychiatry: The Need for Better Data." *European Psychiatry* 42 (May 2017): 134–137. https://doi.org/10.1016/j.eurpsy.2016.12.009.

Drum, Kevin, "Lead: America's Real Criminal Element." *Mother Jones*, February 11, 2016. Accessed August 16, 2018. https://www.motherjones.com/environment/2016/02/lead-exposure-gasoline-crime-increase-children-health/.

Drutman, Lee. "How Corporate Lobbyists Conquered American Democracy." *The Atlantic*, April 20, 2015. Accessed August 16, 2018. https://www.theatlantic.com/business/archive/2015/04/how-corporate-lobbyists-conquered-american-democracy/390822/.

Dwyer, Dustin. "Michigan's Office Reviewing Lawsuit to Hold Companies Liable for Putting Lead in Products." *Michigan Radio*. Accessed July 15, 2019. https://www.michiganradio.org/post/michigan-ags-office-reviewing-lawsuit-hold-companies-liable-putting-lead-products.

Ekong, E. B., B. G. Jaar, and V. M. Weaver. "Lead-Related Nephrotoxicity: A Review of the Epidemiologic Evidence." *Kidney International* 70, no. 12 (December 2, 2006): 2074–2084. https://doi.org/10.1038/sj.ki.5001809.

Eveleth, Rose. "Barns Are Painted Red Because of the Physics of Dying Stars." *Smithsonian Magazine*. Accessed September 19, 2018. https://www.smithsonianmag.com/smart-news/barns-are-painted-red-because-of-the-physics-of-dying-stars-58185724/.

Ewens, Michael, Bryan Tomlin, and Liang Choon Wang. "Statistical Discrimination or Prejudice? A Large Sample Field Experiment." *Review of Economics and Statistics* 96, no. 1 (March 2014): 119–134. https://doi.org/10.1162/REST_a_00365.

Fazel, Seena ,Paul Lichtenstein, Martin Grann, and Niklas Långström. "Risk of Violent Crime in Individuals with Epilepsy and Traumatic Brain Injury: A 35-Year Swedish Population Study." *PLOS Medicine* 8, no. 12 (December 27, 2011): e1001150. http://journals.plos.org/plosmedicine/article?id=10.1371/journal.pmed.1001150.

Federal Aviation Administration. "Aviation Gasoline—About Aviation Gasoline." Accessed July 12, 2019. https://www.faa.gov/about/initiatives/avgas/.

Federal Bureau of Investigation. "Uniform Crime Reporting Statistics." Accessed July 7, 2017. https://www.ucrdatatool.gov/Search/Crime/State/RunCrimeTrendsInOneVar.cfm.

Feigenbaum, James J., and Christopher Muller. "Lead Exposure and Violent Crime in the Early Twentieth Century." *Explorations in Economic History* 62 (October 2016): 51–86. https://doi.org/10.1016/j.eeh.2016.03.002.

Fine, Philip R., Craig W. Thomas, Richard H. Suhs, Rosellen E. Cohnberg, and Bruce A. Flashner. "Pediatric Blood Lead Levels: A Study in 14 Illinois Cities of Intermediate Population." *JAMA* 221, no. 13 (September 25, 1972): 1475–1479. https://doi.org/10.1001/jama.1972.03200260015005.

Flanagan, Daniel M. "Asbestos Data Sheet." Mineral Commodity Summaries 2020. United States Geological Survey, January 2020. https://pubs.usgs.gov/periodicals/mcs2020/mcs2020-asbestos.pdf.

Fortin, Jacey. "Michigan Will No Longer Provide Free Bottled Water to Flint." *New York Times*, April 9, 2018, U.S. sec. https://www.nytimes.com/2018/04/08/us/flint-water-bottles.html.

Fox, James. *Trends in Juvenile Violence: A Report to the United States Attorney General on Current and Future Rates of Juvenile Offending*. Prepared for the Bureau of Justice Statistics, United States Department of Justice, March 1996. https://www.bjs.gov/content/pub/pdf/tjvfox.pdf.

Frankel, Joseph. "The Government Won't Fund Research on Gun Violence Because of NRA Lobbying." *Newsweek*, October 2, 2017. Accessed August 16, 2018. https://www.newsweek.com/government-wont-fund-gun-research-stop-violence-because-nra-lobbying-675794.

Freed, Bruce, and Marian Currinder. "Do Political Business in the Daylight." *US News & World Report*, April 6, 2016. Accessed August 16, 2018. https://www.usnews.com/news/the-report/articles/2016-04-06/corporate-money-is-playing-a-shadowy-role-in-2016-politics.

Free Range Kids (blog). Accessed August 16, 2018. http://www.freerangekids.com/.

Funk, Cary, and Brian Kennedy. "Public Confidence in Scientists Has Remained Stable for Decades." *Pew Research Center* (blog), April 6, 2017. http://www .pewresearch.org/fact-tank/2017/04/06/public-confidence-in-scientists-has -remained-stable-for-decades/.

Furth, Salim. "Fuel Economy Standards Are a Costly Mistake." Heritage Foundation. Accessed August 16, 2018. https://government-regulation/report/fuel-economy -standards-are-costly-mistake.

Gallup, Inc. "More Americans Say Crime Is Rising in U.S." Gallup.com. Accessed August 16, 2018. https://news.gallup.com/poll/186308/americans-say-crime-rising .aspx.

Gao, Hui, Bao-Jun Yang, Nan Li, Li-Min Feng, Xiao-Yu Shi, Wei-Hong Zhao, and Si-Jin Liu. "Bisphenol A and Hormone-Associated Cancers: Current Progress and Perspectives." *Medicine* 94, no. 1 (January 9, 2015). https://doi.org/10.1097/MD .0000000000000211.

Gardner, Howard, and Thomas Hatch. "Multiple Intelligences Go to School: Educational Implications of the Theory of Multiple Intelligences." *Educational Researcher* 18, no. 8 (November 1989): 4. https://doi.org/10.2307/1176460.

Garofalo, James. "NCCD Research Review: Crime and the Mass Media; A Selective Review of Research." *Journal of Research in Crime and Delinquency* 18, no. 2 (July 1, 1981): 319–350. https://doi.org/10.1177/002242788101800207.

Gilbert, Steven G., and Bernard Weiss. "A Rationale for Lowering the Blood Lead Action Level from 10 to 2 μg/dL." *Neurotoxicology* 27, no. 5 (September 2006): 693–701. https://doi.org/10.1016/j.neuro.2006.06.008.

Glaeser, Edward L., and Bruce Sacerdote. "Why Is There More Crime in Cities?" *Journal of Political Economy* 107, no. S6 (December 1, 1999): S225–258. https://doi .org/10.1086/250109.

The Global Risks Report 2018. World Economic Forum. Accessed August 16, 2018. https://www.weforum.org/reports/the-global-risks-report-2018/.

Gneezy, Uri, Stephan Meier, and Pedro Rey-Biel. "When and Why Incentives (Don't) Work to Modify Behavior." *Journal of Economic Perspectives* 25, no. 4 (November 2011): 191–210. https://doi.org/10.1257/jep.25.4.191.

Goldschlag, Nathan, and Alex Tabarrok. "Is Regulation to Blame for the Decline in American Entrepreneurship?" *Economic Policy* 33, no. 93 (January 1, 2018): 5–44. https://doi.org/10.1093/epolic/eix019.

Goodlad, James K., David K. Marcus, and Jessica J. Fulton. "Lead and Attention-Deficit/ Hyperactivity Disorder (ADHD) Symptoms: A Meta-Analysis." *Clinical Psychology Review* 33, no. 3 (April 1, 2013): 417–425. https://doi.org/10.1016/j.cpr.2013.01.009.

Goodman, J. David, and Al Baker. "Murders in New York Drop to a Record Low, but Officers Aren't Celebrating." *New York Times*, January 10, 2018, New York sec. https://www.nytimes.com/2015/01/01/nyregion/new-york-city-murders-fall-but -the-police-arent-celebrating.html.

Goodman, Nathan G. *The Ingenious Dr. Franklin: Selected Scientific Letters of Benjamin Franklin*. Philadelphia: University of Pennsylvania Press, 2011.

Gould, Elise. "Childhood Lead Poisoning: Conservative Estimates of the Social and Economic Benefits of Lead Hazard Control." *Environmental Health Perspectives* 117, no. 7 (July 2009): 1162–1167. https://doi.org/10.1289/ehp.0800408.

Grab, Denise A., and Jack Lienke. "The Falling Cost of Clean Power Plan Compliance." Institute for Policy Integrity, New York University School of Law,

October 2017. http://policyintegrity.org/files/publications/Falling_Cost_of_CPP _Compliance.pdf.

Graham, Judith, and Leslie Forstadt. "Children and Brain Development: What We Know about How Children Learn." Cooperative Extension Publications, University of Maine. Accessed August 15, 2018. https://extension.umaine.edu /publications/4356e/.

Gray, B. H., R. A. Cooke, and A. S. Tannenbaum. "Research Involving Human Subjects." *Science* 201, no. 4361 (September 22, 1978): 1094–1101.

Greenfield Boyce, Nell. "Spending Bill Lets CDC Study Gun Violence; but Researchers Are Skeptical It Will Help." NPR.org, March 23, 2018. Accessed August 16, 2018. https://www.npr.org/sections/health-shots/2018/03/23/596413510/proposed -budget-allows-cdc-to-study-gun-violence-researchers-skeptical.

Grönqvist, Hans, J. Peter Nilsson, and Per-Olof Robling. "Early Lead Exposure and Outcomes in Adulthood." Institute for Evaluation of Labour Market and Education Policy (IFAU), Working Paper No. 2017:4. Uppsala, 2017. http://hdl .handle.net/10419/201482.

Guilarte, Tomás R., Christopher D. Toscano, Jennifer L. McGlothan, and Shelley A. Weaver. "Environmental Enrichment Reverses Cognitive and Molecular Deficits Induced by Developmental Lead Exposure." *Annals of Neurology* 53, no. 1 (2003): 50–56. https://doi.org/10.1002/ana.10399.

Hanna-Attisha, Mona. *What the Eyes Don't See: A Story of Crisis, Resistance, and Hope in an American City.* New York: One World, 2018.

Hanna-Attisha, Mona, Jenny LaChance, Richard Casey Sadler, and Allison Champney Schnepp. "Elevated Blood Lead Levels in Children Associated with the Flint Drinking Water Crisis: A Spatial Analysis of Risk and Public Health Response." *American Journal of Public Health* 106, no. 2 (February 2016): 283–290. https://doi.org/10.2105/AJPH.2015.303003.

Hays, Samuel P. "The Role of Values in Science and Policy: The Case of Lead." In *Explorations in Environmental History,* 291–314. Pittsburgh: University of Pittsburgh Press, 1998.

Health Impact Project. "10 Policies to Prevent and Respond to Childhood Lead Exposure: An Assessment of the Risks Communities Face and Key Federal, State, and Local Solutions," August 2017. https://www.pewtrusts.org/~/media/assets /2017/08/hip_childhood_lead_poisoning_report.pdf.

Heckman, James J., and Paul A. LaFontaine. "The American High School Graduation Rate: Trends and Levels." *Review of Economics and Statistics* 92, no. 2 (May 2010): 244–262. https://doi.org/10.1162/rest.2010.12366.

Heizman, C. W. "Calcium Signaling in the Brain." *Acta Neurobiologiae Experimentalis* 53 (1993): 15.

Henry, Mike. "US R&D Spending at All-Time High, Federal Share Reaches Record Low." *FYI: Science Policy News from the American Institute of Physics,* November 8, 2016. https://www.aip.org/fyi/2016/us-rd-spending-all-time-high-federal-share -reaches-record-low.

Herrnstein, Richard J., and Charles A. Murray. *The Bell Curve: Intelligence and Class Structure in American Life.* New York: Free Press, 1994.

Hershey, Robert D., Jr. "How the Oil Glut Is Changing Business." *New York Times,* June 21, 1981, Business Day sec. https://www.nytimes.com/1981/06/21/business /how-the-oil-glut-is-changing-business.html.

Hillier, Amy E. "Redlining and the Home Owners' Loan Corporation." *Journal of Urban History* 29, no. 4 (2003): 394–420.

Hiltzik, Michael. "Supreme Court Deals Final Blow to Lead Paint Manufacturers' Years-Long Effort to Avoid Cleanup Costs." Latimes.com. Accessed July 15, 2019. https://www.latimes.com/business/hiltzik/la-fi-hiltzik-lead-paint-20181015-story.html.

"Historical Sea Surface Temperature Adjustments/Corrections aka 'The Bucket Model.'" Watts Up with That?, May 25, 2013. https://wattsupwiththat.com/2013/05/25/historical-sea-surface-temperature-adjustmentscorrections-aka-the-bucket-model/.

Hoffman, Kate, Julie A. Sosa, and Heather M. Stapleton. "Do Flame Retardant Chemicals Increase the Risk for Thyroid Dysregulation and Cancer?" *Current Opinion in Oncology* 29, no. 1 (January 2017): 7. https://doi.org/10.1097/CCO.0000000000000335.

Hogan, Bernie, and Brent Berry. "Racial and Ethnic Biases in Rental Housing: An Audit Study of Online Apartment Listings." *City and Community* 10, no. 4 (December 2011): 351–372. https://doi.org/10.1111/j.1540-6040.2011.01376.x.

Holodny, Elena, "Traffic Fatalities in the US Have Been Mostly Plummeting for Decades." *Business Insider*, April 20, 2016. Accessed August 16, 2018. https://www.businessinsider.com/traffic-fatalities-historical-trend-us-2016-4.

"How Much Will the Clean Power Plan Cost?" Union of Concerned Scientists. Accessed August 16, 2018. https://www.ucsusa.org/global-warming/reduce-emissions/how-much-will-clean-power-plan-cost.

"Huntington Williams, Was Health Commissioner." *Baltimore Sun*, May 5, 1992. Accessed August 15, 2018. http://articles.baltimoresun.com/1992-05-05/news/1992126166_1_public-health-hopkins-school-hygiene-and-public.

Hurwitz, Jon, and Mark Peffley. "Explaining the Great Racial Divide: Perceptions of Fairness in the US Criminal Justice System." *Journal of Politics* 67, no. 3 (2005): 762–783.

Iguchi, Taisen, Hajime Watanabe, and Yoshinao Katsu. "Developmental Effects of Estrogenic Agents on Mice, Fish, and Frogs: A Mini-Review." *Hormones and Behavior* 40, no. 2 (September 1, 2001): 248–251. https://doi.org/10.1006/hbeh.2001.1675.

İmrohoroğlu, Ayse, Antonio Merlo, and Peter Rupert. "What Accounts for the Decline in Crime?" *International Economic Review* 45, no. 3 (August 1, 2004): 707–729. https://doi.org/10.1111/j.0020-6598.2004.00284.x.

Innocenti, Giorgio M., and David J. Price. "Exuberance in the Development of Cortical Networks." *Nature Reviews Neuroscience* 6, no. 12 (December 2005): 955–965. https://doi.org/10.1038/nrn1790.

Intergovernmental Panel on Climate Change. "Working Group II: Impacts, Adaptation and Vulnerability." Accessed August 16, 2018. http://www.ipcc.ch/ipccreports/tar/wg2/index.php?idp=29.

Israni, Ellora Thadaney. "When an Algorithm Helps Send You to Prison." *New York Times*, January 20, 2018, Opinion sec. https://www.nytimes.com/2017/10/26/opinion/algorithm-compas-sentencing-bias.html.

Jacobs, David E., Robert P. Clickner, Joey Y. Zhou, Susan M. Viet, David A. Marker, John W. Rogers, Darryl C. Zeldin, Pamela Broene, and Warren Friedman. "The Prevalence of Lead-Based Paint Hazards in U.S. Housing." *Environmental Health Perspectives* 110, no. 10 (October 2002): A599–606.

Jaffe, Adam B., Steven R. Peterson, Paul R. Portney, and Robert N. Stavins.

"Environmental Regulation and the Competitiveness of U.S. Manufacturing: What Does the Evidence Tell Us?" *Journal of Economic Literature* 33, no. 1 (1995): 132–163.

Jee-Lyn García, Jennifer, and Mienah Zulfacar Sharif. "Black Lives Matter: A Commentary on Racism and Public Health." *American Journal of Public Health* 105, no. 8 (August 2015): e27–30. https://doi.org/10.2105/AJPH.2015.302706.

Jensen, Arthur Robert, and Helmuth Nyborg, eds. *The Scientific Study of General Intelligence: Tribute to Arthur R. Jensen.* Amsterdam: Pergamon, 2003.

John of Ephesus. "Ecclesiastical History, Part 3—Book 3." Accessed August 15, 2018. http://www.tertullian.org/fathers/ephesus_3_book3.htm.

Katsiyannis, Antonis, Joseph B. Ryan, Dalun Zhang, and Anastasia Spann. "Juvenile Delinquency and Recidivism: The Impact of Academic Achievement." *Reading and Writing Quarterly* 24, no. 2 (March 4, 2008): 177–196. https://doi.org/10.1080/10573560701808460.

Katz, Diane. "Environmental Policy Guide: 167 Recommendations for Environmental Policy Reform." Heritage Foundation. Accessed August 16, 2018. https://environment/report/environmental-policy-guide-167-recommendations-environmental-policy-reform.

Kehoe, R. A., F. Thamann, and J. Cholak. "On the Normal Absorption and Excretion of Lead. I. Lead Absorption and Excretion in Primitive Life." *Journal of Industrial Hygiene* 15 (1933): 257–272.

Kennedy, Merrit, "Lead-Laced Water in Flint: A Step-by-Step Look at the Makings of a Crisis." NPR.org, April 20, 2016. Accessed August 16, 2018. https://www.npr.org/sections/thetwo-way/2016/04/20/465545378/lead-laced-water-in-flint-a-step-by-step-look-at-the-makings-of-a-crisis.

Kindermann, Charles, James Lynch, and David Cantor. "Effects of the Redesign on Victimization Estimates." U.S. Department of Justice, April 1997. https://www.bjs.gov/content/pub/pdf/ERVE.PDF.

King, Barbara. "Pop Quiz: How Science-Literate Are We, Really?" NPR.org, August 27, 2015. Accessed August 16, 2018. https://www.npr.org/sections/13.7/2015/08/27/435148051/pop-quiz-how-science-literate-are-we-really.

King, Erica. "Black Men Get Longer Prison Sentences Than White Men for the Same Crime: Study." ABC News, November 17, 2017. Accessed August 16, 2018. https://abcnews.go.com/Politics/black-men-sentenced-time-white-men-crime-study/story?id=51203491.

Kirk, Elizabeth A., Jessica Albin, and Rose-Liza Elsma-Osorio, eds. *The Impact of Environmental Law: Stories of the World We Want.* Northampton, MA: Edward Elgar, 2020.

Kitman, Jamie Lincoln. "The Secret History of Lead." *The Nation*, March 2, 2000. https://www.thenation.com/article/secret-history-lead/.

Kolbert, Elizabeth. "Why Facts Don't Change Our Minds." *New Yorker*, February 20, 2017. https://www.newyorker.com/magazine/2017/02/27/why-facts-dont-change-our-minds.

Kordas, Katarzyna, Richard L. Canfield, Patricia López, Jorge L. Rosado, Gonzalo García Vargas, Mariano E. Cebrián, Javier Alatorre Rico, Dolores Ronquillo, and Rebecca J. Stoltzfus. "Deficits in Cognitive Function and Achievement in Mexican First-Graders with Low Blood Lead Concentrations." *Environmental Research* 100, no. 3 (March 1, 2006): 371–386. https://doi.org/10.1016/j.envres.2005.07.007.

Kovarik, William. "Ethyl-Leaded Gasoline: How a Classic Occupational Disease Became an International Public Health Disaster." *International Journal of Occupational and Environmental Health* 11, no. 4 (2005): 384–397. https://doi.org /10.1179/oeh.2005.11.4.384

Krakowski, Menahem. "Violence and Serotonin: Influence of Impulse Control, Affect Regulation, and Social Functioning." *Journal of Neuropsychiatry and Clinical Neurosciences* 15, no. 3 (2003): 294–305.

Kranish, Michael. "Decades-Old Housing Discrimination Case Plagues Donald Trump." NPR.org, September 29, 2016. Accessed August 15, 2018. https://www.npr .org/2016/09/29/495955920/donald-trump-plagued-by-decades-old-housing -discrimination-case.

Kurjak, A., G. Azumendi, N. Veček, S. Kupešic, M. Solak, D. Varga, and F. Cherve-nak. "Fetal Hand Movements and Facial Expression in Normal Pregnancy Studied by Four-Dimensional Sonography." *Journal of Perinatal Medicine* 31, no. 6 (2005): 496–508. https://doi.org/10.1515/JPM.2003.076.

Kurlansky, Mark. "Salt." *New York Times*, February 24, 2002, Books sec. https://www .nytimes.com/2002/02/24/books/chapters/salt.html.

Lafree, Gary, Eric P. Baumer, and Robert O'Brien. "Still Separate and Unequal? A City-Level Analysis of the Black-White Gap in Homicide Arrests since 1960." *American Sociological Review* 71, no. 1 (2010): 75–100. https://doi.org/10.1177 /0003122409357045.

Lahey, Jessica. *The Gift of Failure: How the Best Parents Learn to Let Go so Their Children Can Succeed.* New York: Harper, 2016.

Laidlaw, Mark, Gabriel Filippelli, Richard Sadler, Christopher Gonzales, Andrew Ball, Howard Mielke. "Children's Blood Lead Seasonality in Flint, Michigan (USA), and Soil-Sourced Lead Hazard Risks." *International Journal of Environmental Research and Public Health* 13, no. 4 (March 25, 2016): 358. https://doi.org /10.3390/ijerph13040358.

Lambert, Bruce. "Study Calls LI Most Segregated Suburb." *New York Times*, June 5, 2002.

Landrigan, Philip, Robert Baloh, William Barthel, Randolph Whitworth, Norman Staehling, and Bernard Rosenblum. "Neuropsychological Dysfunction in Children with Chronic Low-Level Lead Absorption." *The Lancet* 305, no. 7909 (1975): 708–712.

Lanphear, Bruce P., Richard Hornung, Jane Khoury, Kimberly Yolton, Peter Baghurst, David C. Bellinger, Richard L. Canfield, et al. "Low-Level Environmental Lead Exposure and Children's Intellectual Function: An International Pooled Analysis." *Environmental Health Perspectives* 113, no. 7 (July 2005): 894–899. https://doi.org /10.1289/ehp.7688.

Lanphear, Bruce P., Stephen Rauch, Peggy Auinger, Ryan W. Allen, and Richard W. Hornung. "Low-Level Lead Exposure and Mortality in US Adults: A Population-Based Cohort Study." *Lancet Public Health* 3, no. 4 (April 1, 2018): e177–184. https://doi.org/10.1016/S2468-2667(18)30025-2.

Lauritsen, Janet L., Maribeth L. Rezey, and Karen Heimer. "When Choice of Data Matters: Analyses of U.S. Crime Trends, 1973–2012." *Journal of Quantitative Criminology* 32, no. 3 (September 1, 2016): 335–355. https://doi.org/10.1007/s10940 -015-9277-2.

"Lead and Healthy Homes Program." City of Philadelphia Public Health. Accessed August 16, 2018. https://www.phila.gov/health/childhoodlead/.

Lee, Michelle Ye Hee. "Does the United States Really Have 5 Percent of the World's Population and One Quarter of the World's Prisoners?" *Washington Post*, April 30,

2015. Accessed August 16, 2018. https://www.washingtonpost.com/news/fact
-checker/wp/2015/04/30/does-the-united-states-really-have-five-percent-of-worlds
-population-and-one-quarter-of-the-worlds-prisoners/.

"LeeAnne Walters." Goldman Environmental Foundation. Accessed August 10, 2018.
https://www.goldmanprize.org/recipient/leeanne-walters/.

Levitt, Steven D. "Understanding Why Crime Fell in the 1990s: Four Factors That
Explain the Decline and Six That Do Not." *Journal of Economic Perspectives* 18,
no. 1 (2004): 163–190.

Li, Wenjie, Shenggao Han, Thomas R. Gregg, Francis W. Kemp, Amy L. Davidow,
Donald B. Louria, Allan Siegel, and John D. Bogden. "Lead Exposure Potentiates
Predatory Attack Behavior in the Cat." *Environmental Research* 92, no. 3 (July 1,
2003): 197–206. https://doi.org/10.1016/S0013-9351(02)00083-X.

Lidsky, T. I., and J. S. Schneider. "Lead Neurotoxicity in Children: Basic Mechanisms
and Clinical Correlates." *Brain* 126, no. 1 (January 1, 2003): 5–19. https://doi.org/10
.1093/brain/awg014.

Little, P., and R. D. Wiffen. "Emission and Deposition of Lead from Motor
Exhausts—II. Airborne Concentration, Particle Size and Deposition of Lead Near
Motorways." *Atmospheric Environment* 12, no. 6 (January 1, 1978): 1331–1341.
https://doi.org/10.1016/0004-6981(78)90073-2.

Loeb, Alan P. "Birth of the Kettering Doctrine: Fordism, Sloanism and the Discovery
of Tetraethyl Lead." *Business and Economic History* 24, no. 1 (Fall 1995): 72–87.

Lovei, Magda. "Phasing Out Lead from Gasoline: Worldwide Experience and Policy
Implications." World Bank Technical Paper No. 397. Washington, DC: World
Bank, 1998.

Lowery, Wesley. "Aren't More White People Than Black People Killed by Police? Yes,
but No." *Washington Post*, July 11, 2016. Accessed August 16, 2018. https://www
.washingtonpost.com/news/post-nation/wp/2016/07/11/arent-more-white-people
-than-black-people-killed-by-police-yes-but-no/?utm_term=.51d39ccb5af6.

Ludwig, J. H., D. R. Diggs, H. E. Hesselberg, and J. A. Maga. "Survey of Lead in the
Atmosphere of Three Urban Communities: A Summary." *American Industrial
Hygiene Association Journal* 26, no. 3 (May 1, 1965): 270–284. https://doi.org/10
.1080/00028896509342731.

Lurie, Julia. "Meet the Mom Who Helped Expose Flint's Toxic Water Nightmare."
Mother Jones, January 21, 2016. Accessed August 10, 2018. https://www.motherjones
.com/politics/2016/01/mother-exposed-flint-lead-contamination-water-crisis/.

Lynch, Jim. "Whistle-Blower Del Toral Grew Tired of EPA 'Cesspool.'" *Detroit News*,
March 28, 2016. Accessed August 10, 2018. https://www.detroitnews.com/story
/news/michigan/flint-water-crisis/2016/03/28/whistle-blower-del-toral-grew-tired
-epa-cesspool/82365470/.

Lythcott-Haims, Julie. *How to Raise an Adult: Break Free of the Overparenting Trap
and Prepare Your Kid for Success.* London: St. Martin's Griffin, 2016.

Marcus, David K., Jessica J. Fulton, and Erin J. Clarke. "Lead and Conduct Problems:
A Meta-Analysis." *Journal of Clinical Child and Adolescent Psychology* 39, no. 2
(February 26, 2010): 234–241. https://doi.org/10.1080/15374411003591455.

Markowitz, Gerald. "The Childhood Lead Poisoning Epidemic in Historical
Perspective." Special issue, *Endeavour; Living in a Toxic World, 1800–2000* 40,
no. 2 (June 1, 2016): 93–101. https://doi.org/10.1016/j.endeavour.2016.03.006.

Markowitz, Gerald E., and David Rosner. *Deceit and Denial: The Deadly Politics of
Industrial Pollution.* Berkeley: University of California Press, 2003.

———. *Lead Wars: The Politics of Science and the Fate of America's Children*. Berkeley: University of California Press; Milbank Memorial Fund, 2014.

Maryon, Herbert. "Metal Working in the Ancient World." *American Journal of Archaeology* 53, no. 2 (1949): 93–125. https://doi.org/10.2307/500498.

Masten, Susan J., Simon H. Davies, and Shawn P. Mcelmurry. "Flint Water Crisis: What Happened and Why?" *Journal—American Water Works Association* 108, no. 12 (December 2016): 22–34. https://doi.org/10.5942/jawwa.2016.108.0195.

Masters, Roger D., Brian Hone, and Anil Doshi. "Environmental Pollution, Neurotoxicity, and Criminal Violence." In *Environmental Toxicology: Current Developments*, edited by J. Rose. London: Taylor and Francis, 1998. http://s3.amazonaws.com/zanran_storage/www.vrfca.org/ContentPages/45257041.pdf.

Matthews, Dylan. "Here's What You Need to Know about Stop and Frisk—and Why the Courts Shut It Down." *Washington Post*, August 13, 2013. Accessed August 16, 2018. https://www.washingtonpost.com/news/wonk/wp/2013/08/13/heres-what-you-need-to-know-about-stop-and-frisk-and-why-the-courts-shut-it-down/.

Mattuck, Rosemary L., Barbara D. Beck, Teresa S. Bowers, and Joshua T. Cohen. "Recent Trends in Childhood Blood Lead Levels." *Archives of Environmental Health* 56, no. 6 (December 11, 2001): 536.

McCarthy, Tom. *Auto Mania: Cars, Consumers, and the Environment*. New Haven, CT: Yale University Press, 2007.

McCoy, Terrence. "Freddie Gray's Life a Study on the Effects of Lead Paint on Poor Blacks." *Washington Post*, April 29, 2015. Accessed July 6, 2017. https://www.washingtonpost.com/local/freddie-grays-life-a-study-in-the-sad-effects-of-lead-paint-on-poor-blacks/2015/04/29/0be898e6-eea8-11e4-8abc-d6aa3bad79dd_story.html?utm_term=.04f23c9b85a5.

McGrayne, Sharon Bertsch. *Prometheans in the Lab: Chemistry and the Making of the Modern World*. New York: McGraw-Hill, 2001.

McNeill, John Robert. *Something New under the Sun: An Environmental History of the Twentieth-Century World*. New York: W. W. Norton, 2000.

McQuaid, John. "Without These Whistleblowers, We May Never Have Known the Full Extent of the Flint Water Crisis." *Smithsonian*, December, 2016. Accessed August 10, 2018. https://www.smithsonianmag.com/innovation/whistleblowers-marc-edwards-and-leeanne-walters-winner-smithsonians-social-progress-ingenuity-award-180961125/.

Meko, Tim, Denise Lu, and Lazaro Gamio. "How Trump Won the Presidency with Razor-Thin Margins in Swing States." *Washington Post*, November 11, 2016. Accessed August 16, 2018. https://www.washingtonpost.com/graphics/politics/2016-election/swing-state-margins/.

Melnick, R. Shep. *Regulation and the Courts: The Case of the Clean Air Act*. Washington, DC: Brookings Institution, 1983.

Mendelsohn, Alan L., Benard P. Dreyer, Arthur H. Fierman, Carolyn M. Rosen, Lori A. Legano, Hillary A. Kruger, Sylvia W. Lim, and Cheryl D. Courtlandt. "Low-Level Lead Exposure and Behavior in Early Childhood." *Pediatrics* 101, no. 3 (March 1, 1998): e10. https://doi.org/10.1542/peds.101.3.e10.

Michigan Civil Rights Commission. "The Flint Water Crisis: Systemic Racism through the Eyes of Flint." February 17, 2017. https://www.michigan.gov/documents/mdcr/VFlintCrisisRep-F-Edited3-13-17_554317_7.pdf.

Mielke, Howard W. "Dynamic Geochemistry of Tetraethyl Lead Dust during the 20th Century: Getting the Lead In, Out, and Translational Beyond." *International*

Journal of Environmental Research and Public Health 15, no. 5 (May 2018): 860–867. https://doi.org/10.3390/ijerph15050860.

———. "Lead in the Inner Cities." *American Scientist* 87 (1999): 62–73.

Mielke, Howard W., Chris R. Gonzales, Eric Powell, Morten Jartun, and Paul W. Mielke. "Nonlinear Association between Soil Lead and Blood Lead of Children in Metropolitan New Orleans, Louisiana: 2000–2005." *Science of the Total Environment* 388, no. 1 (December 15, 2007): 43–53. https://doi.org/10.1016/j.scitotenv.2007.08.012.

Mielke, Howard W., and Sammy Zahran. "The Urban Rise and Fall of Air Lead (Pb) and the Latent Surge and Retreat of Societal Violence." *Environment International* 43 (August 2012): 48–55. https://doi.org/10.1016/j.envint.2012.03.005.

Miller, Earl K., and Jonathan D. Cohen. "An Integrative Theory of Prefrontal Cortex Function." *Annual Review of Neuroscience* 24, no. 1 (2001): 167–202.

Miranda, Marie Lynn, Dohyeong Kim, M. Alicia Overstreet Galeano, Christopher J. Paul, Andrew P. Hull, and S. Philip Morgan. "The Relationship between Early Childhood Blood Lead Levels and Performance on End-of-Grade Tests." *Environmental Health Perspectives* 115, no. 8 (August 2007): 1242–1247. https://doi.org/10.1289/ehp.9994.

Mooney, Chris, and Brady Dennis. "On Climate Change, Scott Pruitt Causes an Uproar—and Contradicts the EPA's Own Website." *Washington Post*, March 9, 2017. Accessed August 16, 2018. https://www.washingtonpost.com/.

Morgenstern, Richard D. *Economic Analyses at EPA: Assessing Regulatory Impact.* Hoboken: Taylor and Francis, 2014. http://public.eblib.com/choice/publicfullrecord.aspx?p=1665806.

Morland, Kimberly B., Susan Filomena, Kathleen Scanlin, James Godbold, Evelyn Granieri, Kelly R. Evenson, Arlene Spark, and Richard Bordowitz. "Neighborhood Environment and Adiposity among Older Adults: The Cardiovascular Health of Seniors and the Built Environment Study." *Medical Research Archives* 5, no. 7 (July 21, 2017). https://www.journals.ke-i.org/index.php/mra/article/view/1361.

Myers, Sage R., Charles C. Branas, Benjamin C. French, Michael L. Nance, Michael J. Kallan, Douglas J. Wiebe, and Brendan G. Carr. "Safety in Numbers: Are Major Cities the Safest Places in the United States?" *Annals of Emergency Medicine* 62, no. 4 (October 1, 2013): 408–418.e3. https://doi.org/10.1016/j.annemergmed.2013.05.030.

"NASA Sea Level Change Portal: Projections." NASA Sea Level Change Portal. Accessed August 16, 2018. https://sealevel.nasa.gov/understanding-sea-level/projections/empirical-projections.

National Center for Environmental Health. "CDC—Lead—Tips—Sources of Lead—Sindoor Alert." Accessed August 16, 2018. https://www.cdc.gov/nceh/lead/tips/sindoor.htm.

———. "CDC—Lead—Tips—Sources of Lead—Toy Jewelry." Accessed August 16, 2018. https://www.cdc.gov/nceh/lead/tips/jewelry.htm.

National Minerals Information Center, United States Geological Survey. "Historical Statistics for Mineral Commodities in the United States, Data Series 2005-140." Accessed August 10, 2018. https://minerals.usgs.gov/minerals/pubs/historical-statistics/#lead.

National Research Council of the National Academies. *Understanding Crime Trends: Workshop Report.* Washington, DC: National Academies Press, 2008.

Naughton, Michael, Frederick Sebold, and Thomas Mayer. "The Impacts of the California Beverage Container Recycling and Litter Reduction Act on

Consumers." *Journal of Consumer Affairs* 24, no. 1 (June 1, 1990): 190–220. https://doi.org/10.1111/j.1745-6606.1990.tb00265.x.

Needleman, Herbert L. "Clamped in a Straitjacket: The Insertion of Lead into Gasoline." *Environmental Research* 74, no. 2 (August 1, 1997): 95–103. https://doi .org/10.1006/enrs.1997.3767.

———. "The Removal of Lead from Gasoline: Historical and Personal Reflections." *Environmental Research* 84, no. 1 (September 2000): 20–35. https://doi.org/10 .1006/enrs.2000.4069.

———. "Salem Comes to the National Institutes of Health: Notes from inside the Crucible of Scientific Integrity." *Pediatrics* 90, no. 6 (December 1, 1992): 977–981.

Needleman, Herbert L., Charles Gunnoe, Alan Leviton, Robert Reed, Henry Peresie, Cornelius Maher, and Peter Barrett. "Deficits in Psychologic and Classroom Performance of Children with Elevated Dentine Lead Levels." *New England Journal of Medicine* 300, no. 13 (March 29, 1979): 689–695. https://doi.org/10.1056 /NEJM197903293001301.

Needleman, Herbert L., Christine McFarland, Roberta B. Ness, Stephen E. Fienberg, and Michael J. Tobin. "Bone Lead Levels in Adjudicated Delinquents: A Case Control Study." *Neurotoxicology and Teratology* 24, no. 6 (2002): 711–717.

Needleman, Herbert L., Julie A. Riess, Michael J. Tobin, Gretchen E. Biesecker, and Joel B. Greenhouse. "Bone Lead Levels and Delinquent Behavior." *JAMA* 275, no. 5 (February 7, 1996): 363–369. https://doi.org/10.1001/jama.1996.03530290033034.

Needleman, Herbert L., Alan Schell, David Bellinger, Alan Leviton, and Elizabeth N. Allred. "The Long-Term Effects of Exposure to Low Doses of Lead in Childhood." *New England Journal of Medicine* 322, no. 2 (January 11, 1990): 83–88. https://doi .org/10.1056/NEJM199001113220203.

Nevin, Rick. "How Lead Exposure Relates to Temporal Changes in IQ, Violent Crime, and Unwed Pregnancy." *Environmental Research* 83, no. 1 (May 1, 2000): 1–22. https://doi.org/10.1006/enrs.1999.4045.

———. *Lucifer Curves*. Pennsauken, NJ: BookBaby, 2016.

———. "Understanding International Crime Trends: The Legacy of Preschool Lead Exposure." *Environmental Research* 104, no. 3 (July 2007): 315–336. https://doi.org /10.1016/j.envres.2007.02.008.

New York State Department of Labor."Labor Statistics for the New York City Region." Accessed June 18, 2020. https://www.labor.ny.gov/stats/nyc/.

Newell, Richard G., and Kristian Rogers. "The U.S. Experience with the Phasedown of Lead in Gasoline." *Resources for the Future*, June 2003. http://web.mit.edu /ckolstad/www/Newell.pdf.

Nigg, Joel T., Molly Nikolas, G. Mark Knottnerus, Kevin Cavanagh, and Karen Friderici. "Confirmation and Extension of Association of Blood Lead with Attention-Deficit/Hyperactivity Disorder (ADHD) and ADHD Symptom Domains at Population-Typical Exposure Levels." *Journal of Child Psychology and Psychiatry, and Allied Disciplines* 51, no. 1 (January 2010): 58–65. https://doi.org/10 .1111/j.1469-7610.2009.02135.x.

"Obama's 'Clean Power Plan' Should Be Called the 'Costly Power Plan.'" *IER* (blog), November 19, 2015. https://www.instituteforenergyresearch.org/fossil-fuels/coal /obamas-clean-power-plan-should-be-called-the-costly-power-plan/.

Oberdörster, Günter, Alison Elder, and Amber Rinderknecht. "Nanoparticles and the Brain: Cause for Concern?" *Journal of Nanoscience and Nanotechnology* 9, no. 8 (August 2009): 4996–5007. https://doi.org/info:doi/10.1166/jnn.2009.GR02.

Oreskes, Naomi, and Erik M. Conway. *Merchants of Doubt: How a Handful of Scientists Obscured the Truth on Issues from Tobacco Smoke to Global Warming.* New York: Bloomsbury Press, 2011.

Ostrander, Rachel R. "School Funding: Inequality in District Funding and the Disparate Impact on Urban and Migrant School Children." *Brigham Young University Education and Law Journal* 1 (2015): 271–295. https://digitalcommons.law.byu.edu/elj/vol2015/iss1/9.

"Our Nation's River: A Troubled Past, a Bright Future." Potomac Conservancy. Accessed August 16, 2018. https://potomac.org/blog/2014/6/27/dc-water-troubled -past-bright-future.

Overby, Peter. "Once Ruled by Washington Insiders, Campaign Finance Reform Goes Grass Roots." NPR.org, April 4, 2016. Accessed August 16, 2018. https://www.npr .org/2016/04/04/473005036/once-ruled-by-washington-insiders-campaign-finance -reform-goes-grassroots.

Paransky, Ora I., and Robert K. Zurawin. "Management of Menstrual Problems and Contraception in Adolescents with Mental Retardation: A Medical, Legal, and Ethical Review with New Suggested Guidelines." *Journal of Pediatric and Adolescent Gynecology* 16, no. 4 (August 2003): 223–235. https://doi.org/10.1016 /S1083-3188(03)00125-6.

Parker, Theodore. "Of Justice and the Conscience." In *Ten Sermons of Religion* (1852). http://www.fusw.org/uploads/1/3/0/4/13041662/of-justice-and-the-conscience.pdf.

Payne, B. Keith, Heidi A. Vuletich, and Kristjen B. Lundberg. "The Bias of Crowds: How Implicit Bias Bridges Personal and Systemic Prejudice." *Psychological Inquiry* 28, no. 4 (October 2, 2017): 233–248. https://doi.org/10.1080/1047840X.2017.1335568.

Philadelphia Childhood Lead Poisoning Prevention Advisory Group. "Final Report and Recommendations." June 20, 2017. https://www.phila.gov/health/pdfs /Lead%20Advisory%20Group%20Report.pdf.

Picchi, Aimee. "The High Price of Incarceration in America." CBS News, May 8, 2014. Accessed August 16, 2018. https://www.cbsnews.com/news/the-high-price-of -americas-incarceration-80-billion/.

Pinker, Steven. *The Better Angels of Our Nature: Why Violence Has Declined.* New York: Viking, 2011.

Pirkle, James L., Debra J. Brody, Elaine W. Gunter, Rachel A. Kramer, Daniel C. Paschal, Katherine M. Flegal, and Thomas D. Matte. "The Decline in Blood Lead Levels in the United States: The National Health and Nutrition Examination Surveys (NHANES)." *JAMA* 272, no. 4 (1994): 284–291.

Plungis, Jeff. "EPA Chief Rejects Obama-Era Fuel Economy Targets." *Consumer Reports*, April 3, 2018. Accessed August 16, 2018. https://www.consumerreports.org /fuel-economy-efficiency/epa-chief-rejects-fuel-economy-standards/.

"Process Matters." Global Automakers. Accessed August 16, 2018. https://www .globalautomakers.org/posts/blog/process-matters.

Rabin, Richard. "The Lead Industry and Lead Water Pipes 'A Modest Campaign.'" *American Journal of Public Health* 98, no. 9 (September 2008): 1584–1592. https://doi.org/10.2105/AJPH.2007.113555.

Rachlinski, Jeffrey J., Sheri Lynn Johnson, Andrew J. Wistrich, and Chris Guthrie. "Does Unconscious Racial Bias Affect Trial Judges?" *Notre Dame Law Review* 84 (2009): 1195–1246.

Raine, Adrian. *The Anatomy of Violence: The Biological Roots of Crime.* New York: Vintage, 2014.

"Reading and Mathematics Score Trends." *The Condition of Education 2016.* National Center for Education Statistics. Accessed August 10, 2018. https://nces.ed.gov/programs/coe/pdf/coe_cnj.pdf.

Rehavi, M. Marit, and Sonja B. Starr. "Racial Disparity in Federal Criminal Sentences." *Journal of Political Economy* 122, no. 6 (December 1, 2014): 1320–1354. https://doi.org/10.1086/677255.

"Report: The War on Marijuana in Black and White." American Civil Liberties Union. Accessed August 16, 2018. https://www.aclu.org/report/report-war-marijuana-black-and-white.

Retief, Francois P., and Louise Cilliers. "Lead Poisoning in Ancient Rome." *Acta Theologica* 26, no. 2 (2006): 147–164.

Reuben, Aaron, Avshalom Caspi, Daniel W. Belsky, Jonathan Broadbent, Honalee Harrington, Karen Sugden, Renate M. Houts, Sandhya Ramrakha, Richie Poulton, and Terrie E. Moffitt. "Association of Childhood Blood Lead Levels with Cognitive Function and Socioeconomic Status at Age 38 Years and with IQ Change and Socioeconomic Mobility between Childhood and Adulthood." *JAMA* 317, no. 12 (March 28, 2017): 1244–1251. https://doi.org/10.1001/jama.2017.1712.

Reyes, Jessica Wolpaw. "Environmental Policy as Social Policy? The Impact of Childhood Lead Exposure on Crime." *BE Journal of Economic Analysis and Policy* 7, no. 1 (2007): 51.

———. "Lead Policy and Academic Performance: Insights from Massachusetts." *Harvard Educational Review* 85, no. 1 (March 18, 2015): 75–107. https://doi.org/10.17763/haer.85.1.bj34u74714022730.

Rhoten, D. "Risks and Rewards of an Interdisciplinary Research Path." *Science* 306, no. 5704 (December 17, 2004): 2046. https://doi.org/10.1126/science.1103628.

Ritter, Steve. "Pencils & Pencil Lead." *Chemical & Engineering News* 79, no. 35. (2001). 10.1021/cen-v079n042.p035.

Robbins, Norman, Zhong-Fa Zhang, Jiayang Sun, Michael E. Ketterer, James A. Lalumandier, and Richard A. Shulze. "Childhood Lead Exposure and Uptake in Teeth in the Cleveland Area during the Era of Leaded Gasoline." *Science of the Total Environment* 408, no. 19 (September 1, 2010): 4118–4127. https://doi.org/10.1016/j.scitotenv.2010.04.060.

Rodríguez-Barranco, Miguel, Marina Lacasaña, Clemente Aguilar-Garduño, Juan Alguacil, Fernando Gil, Beatriz González-Alzaga, and Antonio Rojas-García. "Association of Arsenic, Cadmium and Manganese Exposure with Neurodevelopment and Behavioural Disorders in Children: A Systematic Review and Meta-Analysis." *Science of the Total Environment* 454–455 (June 1, 2013): 562–577. https://doi.org/10.1016/j.scitotenv.2013.03.047.

Roeder, Oliver, Lauren-Brooke Eisen, and Julia Bowling. "What Caused the Crime Decline?" Brennan Center for Justice at NYU School of Law, 2015. https://www.brennancenter.org/sites/default/files/analysis/What_Caused_The_Crime_Decline.pdf.

Roosevelt, Franklin. "Toll Roads and Free Roads." 76th Congress, 1st Session, House Document no. 272, April 27, 1939. http://www.virginiaplaces.org/transportation/tollroadsfreeroads.pdf.

Rosner, David, and Gerald Markowitz. *Dying for Work: Workers' Safety and Health in Twentieth-Century America.* Bloomington: Indiana University Press, 1989.

———. "A 'Gift of God'? The Public Health Controversy over Leaded Gasoline during the 1920s." *American Journal of Public Health* 75, no. 4 (1985): 344–352.

Ross, Martin, David Hoppock, and Brian Murray. "Ongoing Evolution of the Electricity Industry: Effects of Market Conditions and the Clean Power Plan on States." NI WP 16-07. Durham, NC: Duke University, July 21, 2016. https:// nicholasinstitute.duke.edu/climate/publications/ongoing-evolution-electricity -industry-effects-market-conditions-and-clean-power-plan.

Rosselli, Mónica, and Alfredo Ardila. "The Impact of Culture and Education on Non-verbal Neuropsychological Measurements: A Critical Review." *Brain and Cognition* 52, no. 3 (August 2003): 326–333. https://doi.org/10.1016/S0278 -2626(03)00170-2.

Rothstein, Richard. *The Color of Law: A Forgotten History of How Our Government Segregated America*. New York: Liveright, 2017.

Roy, Siddhartha. "Commentary: MDEQ Mistakes and Deception Created the Flint Water Crisis." *Flint Water Study Updates* (blog), September 30, 2015. http:// flintwaterstudy.org/2015/09/commentary-mdeq-mistakes-deception-flint-water -crisis/.

———. "Our Sampling of 252 Homes Demonstrates a High Lead in Water Risk: Flint Should Be Failing to Meet the EPA Lead and Copper Rule." *Flint Water Study Updates* (blog), September 8, 2015. http://flintwaterstudy.org/2015/09/our -sampling-of-252-homes-demonstrates-a-high-lead-in-water-risk-flint-should-be -failing-to-meet-the-epa-lead-and-copper-rule/.

Ruderman, Wendy, Barbara Laker, and Dylan Purcell. "In Booming Philly Neighbor-hoods, Lead-Poisoned Soil Is Resurfacing." *Philadelphia Inquirer*, June 18, 2017. Accessed August 16, 2018. http://www.philly.com/philly/news/special_packages /toxic-city/philadelphia-lead-soil-fishtown-construction-dust.html.

Sanders, Talia, Yiming Liu, Virginia Buchner, and Paul B. Tchounwou. "Neurotoxic Effects and Biomarkers of Lead Exposure: A Review." *Reviews on Environmental Health* 24, no. 1 (March 2009): 15.

Schneider, Stephen H. *Science as a Contact Sport: Inside the Battle to Save Earth's Climate*. Washington, DC: National Geographic, 2009.

Schwartz, Joel, Jane Leggett, Bart Ostro, Hugh Pitcher, and Ronnie Levin. "Costs and Benefits of Reducing Lead in Gasoline." Environmental Protection Agency, March 26, 1984. https://nepis.epa.gov/Exe/ZyPURL.cgi?Dockey=9101KHJA .TXT

Scialla, Mark. "It Could Take Centuries for EPA to Test All the Unregulated Chemicals under a New Landmark Bill." *PBS NewsHour*, June 22, 2016. https:// www.pbs.org/newshour/science/it-could-take-centuries-for-epa-to-test-all-the -unregulated-chemicals-under-a-new-landmark-bill.

"Selecting the Right Octane Fuel." U.S. Department of Energy Office of Renewable Energy and Energy Efficiency. Accessed August 15, 2018. http://www.fueleconomy .gov/feg/octane.shtml.

Sharkey, Patrick. *Uneasy Peace: The Great Crime Decline, the Renewal of City Life, and the Next War on Violence*. 1st ed. New York: W. W. Norton, 2018.

Sharkey, Patrick, Gerard Torrats-Espinosa, and Delaram Takyar. "Community and the Crime Decline: The Causal Effect of Local Nonprofits on Violent Crime." *American Sociological Review* 82, no. 6 (December 1, 2017): 1214–1240. https://doi .org/10.1177/0003122417736289.

Shavit, Elinoar, and Efrat Shavit. "Lead and Arsenic in Morchella Esculenta Fruitbod-ies Collected in Lead Arsenate Contaminated Apple Orchards in the Northeastern United States: A Preliminary Study." *Fungi Magazine* 3, no. 2 (Spring 2010): 11–18.

Sinclair, Tara M., and Kathryn Vesey. "Regulation, Jobs, and Economic Growth: An Empirical Analysis." George Washington University Regulatory Studies Center Working Paper, March 2012. https://pdfs.semanticscholar.org/305f/420d5f36f169c7e2e3b5a512ffae9834fada.pdf.

Sinclair, Upton. *I, Candidate for Governor: And How I Got Licked.* Berkeley: University of California Press, 1994.

Snaidero, Nicolas, and Mikael Simons. "Myelination at a Glance." *Journal of Cell Science* 127, no. 14 (July 15, 2014): 2999–3004. https://doi.org/10.1242/jcs.151043.

Snyder, Howard N. "Arrest in the United States, 1990–2010." U.S. Department of Justice, Office of Justice Programs Bureau of Justice Statistics, October 2012. https://www.bjs.gov/content/pub/pdf/aus9010.pdf.

Sohn, Emily. "Lead: Versatile Metal, Long Legacy." Dartmouth Toxic Metals Superfund Research Program. Accessed August 15, 2018. https://www.dartmouth.edu/~toxmetal/toxic-metals/more-metals/lead-history.html.

Stiles, Joan, and Terry L. Jernigan. "The Basics of Brain Development." *Neuropsychology Review* 20, no. 4 (December 2010): 327–348. https://doi.org/10.1007/s11065-010-9148-4.

"Stop-and-Frisk Data." New York Civil Liberties Union, January 2, 2012. https://www.nyclu.org/en/stop-and-frisk-data%20.

Sumner, Thomas. "ScienceShot: Did Lead Poisoning Bring Down Ancient Rome?" *Science*, April 21, 2014. http://www.sciencemag.org/news/2014/04/scienceshot-did-lead-poisoning-bring-down-ancient-rome.

Taibbi, Matt. "The Shame of Three Strikes Laws." *Rolling Stone* (blog), March 27, 2013. https://www.rollingstone.com/politics/politics-news/cruel-and-unusual-punishment-the-shame-of-three-strikes-laws-92042/.

Taylor, Mark Patrick, Miriam K. Forbes, Brian Opeskin, Nick Parr, and Bruce P. Lanphear. "The Relationship between Atmospheric Lead Emissions and Aggressive Crime: An Ecological Study." *Environmental Health* 15 (February 16, 2016): 23. https://doi.org/10.1186/s12940-016-0122-3.

Tcherni-Buzzeo, Maria. "The 'Great American Crime Decline': Possible Explanations." In *Handbook on Crime and Deviance*, edited by Marvin D. Krohn, Nicole Hendrix, Gina Penly Hall, and Alan J. Lizotte, 309–335. Switzerland: Springer International Publishing, 2019. https://doi.org/10.1007/978-3-030-20779-3_16.

Tonachel, Luke. "2025 Clean Car Standards Are Achievable, Study Shows." Natural Resources Defense Council. Accessed August 16, 2018. https://www.nrdc.org/experts/luke-tonachel/2025-clean-car-standards-are-achievable-study-shows.

Trudeau. Garry. *Doonesbury*, January 28, 1982. GoComics. Accessed August 15, 2018. https://www.gocomics.com/doonesbury/1982/01/28.

Tuthill, Robert W. "Hair Lead Levels Related to Children's Classroom Attention-Deficit Behavior." *Archives of Environmental Health: An International Journal* 51, no. 3 (June 1, 1996): 214–220. https://doi.org/10.1080/00039896.1996.9936018.

US Census Bureau. "Estimates of U.S. Population by Age and Sex." United States Census Bureau. Accessed August 16, 2018. https://www.census.gov/newsroom/press-releases/2018/pop-characteristics.html.

US EPA. "Air Quality—National Summary." Data and Tools. US EPA, May 4, 2016. https://www.epa.gov/air-trends/air-quality-national-summary.

———. "EPA Takes Final Step in Phaseout of Leaded Gasoline." Press release, January 29, 1996. Accessed August 15, 2018. https://archive.epa.gov/epa/aboutepa/epa-requires-phase-out-lead-all-grades-gasoline.html

———. "Fast Facts on Transportation Greenhouse Gas Emissions." Overviews and Factsheets. US EPA, August 25, 2015. https://www.epa.gov/greenvehicles/fast-facts-transportation-greenhouse-gas-emissions.

———. "History of Reducing Air Pollution from Transportation in the United States." Overviews and Factsheets. US EPA, September 10, 2015. https://www.epa.gov/transportation-air-pollution-and-climate-change/accomplishments-and-success-air-pollution-transportation.

———. "Light-Duty Vehicle Greenhouse Gas Emission Standards and Corporate Average Fuel Economy Standards: EPA Response to Comments Document for Joint Rulemaking." Accessed August 16, 2018. https://nepis.epa.gov/Exe/ZyPURL.cgi?Dockey=P1006VMH.TXT.

———. "Real Estate Disclosure." Overviews and Factsheets. US EPA, February 12, 2013. https://www.epa.gov/lead/real-estate-disclosure.

———. "Renovation, Repair and Painting Program." Collections and Lists. US EPA, February 12, 2013. https://www.epa.gov/lead/renovation-repair-and-painting-program.

US EPA, OECA. "Q&A: Orphan Share Superfund Reform." Policies and Guidance. US EPA, October 31, 2013. https://www.epa.gov/enforcement/qa-orphan-share-superfund-reform.

"U.S. Teen Pregnancy, Birth and Abortion Rates Reach Historic Lows." Guttmacher Institute, May 5, 2014. https://www.guttmacher.org/news-release/2014/us-teen-pregnancy-birth-and-abortion-rates-reach-historic-lows.

Vaccari, David A. "How Not to Get the Lead Out—Lead Service Line Replacement Will Not Solve Our Drinking Water Crisis." *Current Pollution Reports* 2, no. 3 (September 1, 2016): 200–202. https://doi.org/10.1007/s40726-016-0034-4.

Van Ulirsch, Gregory, Kevin Gleason, Shawn Gerstenberger, Daphne B. Moffett, Glenn Pulliam, Tariq Ahmed, and Jerald Fagliano. "Evaluating and Regulating Lead in Synthetic Turf." *Environmental Health Perspectives* 118, no. 10 (October 2010): 1345–1349. https://doi.org/10.1289/ehp.1002239.

Vigen, Tyler. *Spurious Correlations*. New York: Hachette Books, 2015.

Wang, Chao-Ling, Hung-Yi Chuang, Chi-Kung Ho, Chun-Yuh Yang, Jin-Lian Tsai, Ting-Shan Wu, and Trong-Neng Wu. "Relationship between Blood Lead Concentrations and Learning Achievement among Primary School Children in Taiwan." *Environmental Research* 89, no. 1 (May 1, 2002): 12–18. https://doi.org/10.1006/enrs.2002.4342.

Warren, Christian. *Brush with Death: A Social History of Lead Poisoning*. Baltimore: Johns Hopkins University Press, 2000.

Weisburd, David, Joshua C. Hinkle, Anthony A. Braga, and Alese Wooditch. "Understanding the Mechanisms Underlying Broken Windows Policing: The Need for Evaluation Evidence." *Journal of Research in Crime and Delinquency* 52, no. 4 (July 1, 2015): 589–608. https://doi.org/10.1177/0022427815577837.

Weiss, Marissa. "Is Acid Rain a Thing of the Past?" *Science*, June 28, 2012. http://www.sciencemag.org/news/2012/06/acid-rain-thing-past.

"When Our Rivers Caught Fire." Michigan Environmental Council, July 11, 2011. Accessed August 16, 2018. https://www.environmentalcouncil.org/when_our_rivers_caught_fire.

Williams, Casey. "Driving the Ford Model T through 110 Years of American Audacity." *Chicago Tribune*, July 3, 2018. Accessed August 15, 2018. http://www.chicagotribune.com/classified/automotive/sc-auto-cover-0705-ford-model-t-20180629-story.html.

Williams, Huntington, Emanuel Kaplan, Charles E. Couchman, and R. R. Sayers. "Lead Poisoning in Young Children." *Public Health Reports* 67, no. 3 (March 1952): 230–236.

Winder, C. "Lead, Reproduction and Development." *Neurotoxicology* 14, no. 2–3 (1993): 303–317.

Winerip, Michael, Michael Schwirtz, and Robert Gebeloff. "For Blacks Facing Parole in New York State, Signs of a Broken System." *New York Times*, January 20, 2018, New York sec. https://www.nytimes.com/2016/12/04/nyregion/new-york-prisons-inmates-parole-race.html.

Wright, John Paul, Kim N. Dietrich, M. Douglas Ris, Richard W. Hornung, Stephanie D. Wessel, Bruce P. Lanphear, Mona Ho, and Mary N. Rae. "Association of Prenatal and Childhood Blood Lead Concentrations with Criminal Arrests in Early Adulthood." *PLOS Medicine* 5, no. 5 (May 27, 2008): e101. https://doi.org/10.1371/journal.pmed.0050101.

Yang, Jia Lynn. "Does Government Regulation Really Kill Jobs? Economists Say Overall Effect Is Minimal." *Washington Post*, November 13, 2011. Accessed August 16, 2018. https://www.washingtonpost.com/business/economy/does-government-regulation-really-kill-jobs-economists-say-overall-effect-minimal/2011/10/19/gIQALRF5IN_story.html.

Zimring, Franklin E., *The Great American Crime Decline*. New York: Oxford University Press, 2008.

Zimring, Franklin E., and Gordon Hawkins. *Crime Is Not the Problem: Lethal Violence in America*. New York: Oxford University Press, 1997.

Zimring, Franklin E., Sam Kamin, and Gordon Hawkins, *Crime and Punishment in California: The Impact of Three Strikes and You're Out*. Berkeley: Institute of Governmental Studies Press, University of California, 1999. https://www.ncjrs.gov/App/Publications/abstract.aspx?ID=183617.

Index

Note: Page numbers for figures and diagrams are in italics.

About the Author

CARRIE NIELSEN received her undergraduate degree in environmental science from Brown University, where she had a work-study job testing lead levels in soil and drinking water from low-income neighborhoods around Providence, Rhode Island. Her PhD in geological and environmental sciences from Stanford University focused on nutrient cycling by bacteria and fungi in tropical forest soils. She is now associate professor of biology and environmental science at Cabrini University, where her research interests have included science pedagogy, interdisciplinary teaching, watershed management, and faith perspectives on environmental sustainability. Carrie lives in Bryn Mawr, Pennsylvania, with her husband, Bill, a playwright and marketing director for the Wilma Theater in Philadelphia, and her two Generation Z daughters, Celia and Anya. This is Carrie's first book, though she did once write a play for young audiences, *Rainbowtown*, which was performed as part of the 2014 Philadelphia FringeArts festival.

Printed and bound by CPI Group (UK) Ltd, Croydon, CR0 4YY

27/10/2024

14580230-0005